HAVE YOU ANY WOOL?

1st Edition

Published in 2013 by
Woodfield Publishing Ltd
Bognor Regis PO21 5EL England
www.woodfieldpublishing.co.uk

Copyright © 2013 Eileen Sullivan

ISBN 1-84683-150-4

Printed and bound in England

Have You Any Wool?

Reflections on a rural life

EILEEN SULLIVAN

Woodfield

Woodfield Publishing Ltd

Bognor Regis ~ West Sussex ~ England ~ PO21 5EL
tel 01243 821234 ~ **e/m** info@woodfieldpublishing.co.uk

Interesting and informative books on a variety of subjects

For full details of all our published titles, visit our website at
www.woodfieldpublishing.co.uk

To our children
and grandchildren

~ CONTENTS ~

Preface

In this book I hope to fill in the gaps missing in *The Sheep's in the Meadow*. I've included more about David's early life and also my childhood . Our married life started off with having to cope with living a hundred miles from Aldershot, which was where our homes were. I didn't know anybody and had to learn the job of being a farmer's wife. My neighbour's husband worked on the farm and his wife worked part time but eventually we became friends – discussing things like whether to use castor sugar or granulated in cakes. Mrs Shorter had a ten year old daughter, Shirley, whom I sometimes cared for in the school holidays. David and I relied on our bikes for getting around. Ashford was eight miles away and I sometimes cycled there if I had an appointment with the dentist.

Living in our next house in rural Sussex was different altogether, with no neighbours at all; the nearest was a mile away. The house was big and rambling with a garden to match, which we never did master. David gained further qualifications, which helped to pave the way when he needed another job.

Our next move was to Hurstbourne Tarrant in Hampshire and we still live in the area. David is self-employed but working shorter hours than he used to. We found a place to worship at the Gospel Hall at Vernham Dean, about four miles away. We still attend there and continue to help with "Friday Club," for Primary age children where we have an average of sixteen children each week. Andrew, our elder son helps with a similar church at Shipton Bellinger. Our other son, Peter attends the church in Aldershot where we worshipped before we were married.

Over the years since David became self-employed, he's had varying flock sizes and at present has forty of his own ewes and five hundred and fifty "lodgers." These sheep come from Kent and David has them as lambs for a year until they are ready for breeding.

David is still in demand for public speaking at events as diverse as 'The Romsey Ramblers Club' and as the after dinner speaker at the John Edgar Memorial Trust annual dinner.

I trust that you will enjoy *Have You Any Wool*, which has more stories, mainly about David and of course the animals in our life.

It is now five years since the publication of *The Sheep's in the Meadow* and I am grateful for the interest shown and the number of books that have been sold.

I hope you enjoy this new book and find in it something to make you smile.

Eileen Sullivan, March 2013

Foreword

Eileen and David Sullivan are a remarkable couple whose energy and enthusiasm for life is nothing short of inspirational. In a long and eventful life together they have worked as a team wherever Dave's agricultural career has led him, raised a family, looked after a variety of animals, including many thousands of sheep – and still found time to volunteer for a variety of activities within the local community, particularly focusing on providing entertainment and guidance for the young.

They recently celebrated their Golden Wedding anniversary and in this enjoyable follow-up to her first and very well received book (*The Sheep's In The Meadow... Hopefully*, which Woodfield had the pleasure of publishing some years ago), Eileen looks back on their many eventful years together, mostly spent in rural Hampshire, where they still live.

For most of his working life, Dave has been involved with sheep farming in a variety of capacities, firstly working for other farmers and later for himself. This book provides plenty of interesting details about lambing, shearing, sheepdogs and much else that will give the uninitiated a good insight into what is involved in sheep husbandry.

In addition to their agricultural working life, over many years Eileen and David have also been involved as volunteers in organising activities for young people, including summer camps, Sunday school, trailer rides and much else. David is also highly regarded as a lay preacher and after-dinner speaker and has made numerous appearances in both capacities over the years, in spite of the demands of his hectic working life and other voluntary commitments.

Of course, David and Eileen's married life has not been without its fair share of difficulties and setbacks, from unexpected illnesses and injuries, to the difficulties of coping with the needs of ageing relatives, to the challenges of raising a family – all of which are discussed in a refreshingly forthright and down-to-earth manner. I'm sure that readers will be able to draw comparisons with similar situations from their own lives and perhaps draw inspiration from the positive manner in which David and Eileen have dealt with the adverse circumstances that have come their way from time to time.

We are happy to publish this second volume of Eileen's memoirs and are confident that readers of all ages and persuasions will find much in it to enjoy.

Nick Shepperd
Editor/Director
Woodfield Publishing Ltd

1.   The War Years

"Don't let Eileen get near the mangle!"

"All right Mum," chorused Pam and Joyce.

Mum had given in to the request of Pam and Joyce to do some dolls washing. The mangle in question was a formidable piece of machinery with big, heavy, wooden rollers.

Mum found some old bits of soap, filled a galvanized bowl with warm water, and carried it through to the shed. With the repeated admonition, "Don't let Eileen near the mangle," she went back indoors.

We fetched our dolls and stripped them off; mine was one with a crock head and a black, stuffed body, her name was Jennifer. She was wearing a pretty blue dress and knickers and I thought she was lovely.

We all began scrubbing the various items. We squeezed the water out and then came the fun part – using the mangle. Joyce took on the job of turning the handle, which was quite hard, and Pam fed the clothes into it. I was supposed to be out of the way. I did not want to miss the fun so took my dolly's clothes over to the mangle.

"Get out of the way Eileen. You know Mum said that you're not to go near the mangle," Pam said, as she picked up a dress that had just gone through the rollers. Joyce didn't see me and carried on turning the handle as I put my dolls clothes up on the bottom roller. She carried on turning the handle and the dress was pulled away by the rollers – as were my fingers. I started screaming, which brought Mum running and made Pam and Joyce cry as Mum shouted at them. My fingers looked – mangled! It was three o'clock in the afternoon, there was no one to look after the girls so Mum had to take us all with her on the bus to the Casualty Department at the hospital in Aldershot.

It was decided that there were no bones broken and the swelling would go down in a few days. As a reward for not crying when my fingers were examined, I was allowed to put sixpence in the charity box at the hospital, which was shaped like the hospital! I was three years old at the time.

Mum and Dad had 'itchy feet' and moved every few years. Before the age of five, I moved four times; from Ash to Ashford and Feltham and then back to Ash in the Aldershot area again.

During the war, we stayed put for about six years as Mum and Dad had obtained a council house for 'the duration'. I had no idea what those words meant until the end of the war when the house had to revert to the original dweller who had been living in Scotland for 'the duration', so once again we had to find somewhere to live. Our whole life seemed geared to the words – 'we moved house again.'

The war years were punctuated with what seemed an endless stream of relations who had been bombed out of their London houses. We only had a two bedroomed house and although Dad was away, there were three of us girls. We seemed to end up sleeping on 'the chair bed' or the ottoman as, invariably, our beds were occupied by aunts and cousins. We lived on a council estate with half the houses occupied by Romany gypsies, which created a 'them' and 'us' mentality. Mum took in needlework to help with the money, and became very ingenious transforming dresses into skirts; skirts into trousers or vice versa and her specialty –'pinners' or pinafores for the gypsies, which were made from blackout material. The government supplied this to make curtains to cover the windows so that light could not be seen outside. Mum hated asking for money so we girls would have the job of taking the altered garments back and asking for payment. We ourselves were clothed mostly in cast offs and Mum was an avid knitter, unpicking cardigans that were past their best and making jerseys into 'pixie' hoods, or scarves for us. As we got older, she taught us to knit. We started with scarves for our dolls and as we became more proficient graduated to striped scarves for ourselves. A lot of tugging and stretching went on before we thought that they were long enough.

"Mum, please have you any wool?" became our constant request and the smallest length would be incorporated to our pathetic efforts at knitting.

Dad was in the Tank Corps at Bovington; his hearing had always been bad – we were told that as a child Grandma had poured neat hydrogen peroxide into his ears to clean them. Added to the general stress, he had become stone deaf and was working as a repair and alterations tailor in the army. It had been his job in 'Civvy Street' but I always wondered if he had really wanted to drive tanks. He would come home on leave sometimes but seemed a very morose man. Mum and Dad loved light classical music and as we had an old piano, Dad could manage, after a fashion, to play by ear Tchaikovsky or Rachmaninov piano concertos, Debussy's Claire de Lune and similar melodies.

The highlight of his leaves was usually a visit to 'The Hippodrome' in Aldershot to a variety show or the Theatre Royal to see a play. There were also about six cinemas from which to choose and we used to go to Saturday morning pictures at the Empire; the queue to get in would stretch right round the building. At the beginning of the war there would still be an organist rising up from the floor on his mighty 'Wurlitzer' organ (which kept changing colour) to play songs that we could all join in. Some of the songs were popular in the First World War – 'Roll out the barrel', 'It's a long way to Tipperary' 'My old man' – but we roared them out with everyone else. Some of them were quite sentimental; 'There'll be bluebirds over the white cliffs of Dover' was my favourite. We loved the films that featured 'Old Mother Riley' and her 'daughter' Kitty. I could never understand how Old Mother Riley with her old clothes and hair in a bun could really be a man, and, more amazing that her 'daughter' Kitty was really her 'wife!'

Even when Dad wasn't around to take us anywhere, we were still able to go occasionally to an entertainment. Joyce, my eldest sister was seven years older than I was and Pam was four years older than me. I can remember trying to recall crude jokes we heard at the Hippodrome to tell Mum but Joyce made me promise not to tell her. I was also informed that the tableaux of nudes really had flesh coloured 'body suits' and again came the warning don't tell Mum! At the end of the performance we would all stand in silence while the National Anthem was played then run down the stairs from the 'gods', hoping that we hadn't lost the bus home. In spite of the heavy presence of the military, we never felt intimidated.

We were fascinated by soldiers who were recuperating in the Cambridge Army Hospital, on the edge of the army camp. Those mobile enough to go out could be seen in Aldershot in their distinctive uniform of khaki battledress and bright blue trousers, some with empty sleeves pinned across their chests. The other thing that brought the war close to us in the villages was the convoys of army vehicles travelling through the village to the railway station. We would often have to stand and watch them pass while we were waiting to cross the road after school. The tanks fascinated us especially the big ones and we tried to imagine what it was like to be inside one with no windows and the tremendous noise they made.

The local Church 'Guild' would sometimes put on a concert in the church hall in aid of the Red Cross or similar organizations. It was always performed by the ladies although sometimes the piano was played by a male retired musician. A soloist would sing a couple of patriotic songs, 'Hearts of oak' was a favourite or sometimes a wavering aged tenor would attempt, 'Come into the garden Maud'. It is the only place I ever saw a performance by a 'musical saw' artist; he used a violin bow and bent the saw for the various notes making a unique sound.

A patriotic sketch would be performed with ladies taking the parts of men, unconvincingly dressed in uniforms or jacket and trousers, with an unlit pipe or cigarette in their mouths. This was heartily applauded; and so the evening moved on culminating with the audience joining in with singing 'Jerusalem' or 'Rule Britannia' and of course the National Anthem. A 'retiring' collection would be taken and we would all go home in the dark, no torches because of the blackout regulations.

There were a lot of 'jingles' we sang about the war, although at the time I had no idea what we were singing about. One, which went to the tune of 'Whistle while you work' went:

> *'Whistle while you work,*
> *Mussolini is a twerp,*
> *Goering's barmy,*
> *So's 'is army,*
> *Rub them in the dirt.'*

My sisters never let me forget a question I then asked.

"Who's 'Sosis?" As far as I was concerned he had an army! Another one was:

'Heil, Hitler, Yah, yah, yah,
O what a funny little man you are,
With your little moustache and your hair all 'blosh'
Heil Hitler, yah,yah, yah.'

In the summer, we used to spend a lot of time at Aldershot Bathing Pool (now renamed Aldershot Lido and frequented by my grandchildren, who live in the town). A gang of us from the estate used to go and spend the day there, taking sandwiches, a biscuit and a bottle of lime juice, which I can't remember drinking anywhere else. We used to walk the mile and a half to get there or sometimes, if Mum had enough money, catch the bus for half of the way. I still can't swim very well but I enjoy trying. We would get undressed in a communal room known as 'the cowshed' because that's what it looked like – to use the changing rooms cost money!

Anyone who used Aldershot Bathing Pool then will agree that it was an unforgettable place. It had much larger grounds then than it has now plus a 'stream' running all the way round, which seemed a long way then. We loved to watch the men diving, as the facilities were good enough for the Olympic Games to be held there when England hosted them in 1948. The chute fascinated us and I longed to come whooshing down it but I never had the courage to try. There was a fountain in the middle (which is still there) and a board, from which races were started. My grandchildren can all swim well and spend a lot of their summer holidays enjoying the same delights that we did. Dave lived near enough to cycle or walk to it and had a season ticket each year but at that time we hadn't met. It was drained during the war because it would be a landmark for the German bombers to aim for.

One of our other delights was to go for walks across Ash Common, which was a vast area of scrubland belonging to the army. We knew that if the red flags were flying we were to keep away from the firing ranges. We found a network of army trenches dug into the sandy soil; it made a lovely labyrinth in which to play hide-and-seek with the added attraction that we knew we shouldn't be there.

In the autumn we used to collect pockets full of sweet chestnuts; it was the only place locally where they grew. We would put them on the fire to roast when we got home and it's a taste that is hard to forget. One of our favourite spots was the Fox Hills – a high place with a fantastic view – and an army pillbox that we would appropriate for our games; there would usually be about ten of us. There were huge army ants scurrying around and we had to keep a look out for snakes.

With money short, we managed somehow, as Mum had her needlework right through the war. Coal became scarce but Mum heard that there was coke to be had at Aldershot Gas Works, a mile and a half away. Somebody had an old pram so she borrowed that and Joyce, Pam and I were sent to get two shilling's worth of coke. To get to the gas works we had to pass the slaughterhouse. I'm not sure, which smelled worse!

We managed to find the right place and the man tipped up the sack into the pram, which was now quite heavy to push. We parted with the two shillings and set off home. It took us quite a while, as the combined strength of Pam and Joyce was needed. I was just a nuisance, as I was tired and dawdled behind, grizzling. Mum was pleased to see us back and she shovelled the coke into the coal bunker, leaving some in the pram for a 'thank you' to the lady who owned the pram.

'Mum, can we go up the end with Sheila and Theresa?' Joyce queried.

'All right but be back before tea as there's some errands to run – and mind you look after Eileen!'

'Thanks Mum.'

We knew that something exciting was happening on the waste piece of ground at the end of the road. Other children were heading in the same direction. The land belonged to the council, I think; the gypsies were allowed to keep their horses on it and it was a lovely place from which to fly our homemade kites – made with brown paper, twigs and string if we could get it – or to build dens. The interest seemed to be at the far side so we joined the crowd, which was gathered round something lying on the ground; it was a horse. It had its head thrown back and was absolutely still and we knew without being told that it was dead. We hadn't seen anything dead before; when the cat had to be put to sleep we just left it with the slaughterhouse man, together with the two shillings he asked for doing the job.

We tried to find out what was going to happen to the horse, as digging a grave didn't seem possible for such a large animal. Then we caught a whisper, "The knacker man is coming." We had no idea what a knacker man did but we soon found out. He drove up in a battered old lorry with some machinery in the back and reversed up to the horse. We jostled each other to get a better place from which to see. He pulled a chain from the back of the lorry and secured it around the horse's neck; he then operated a switch, which lowered the floor of the lorry, then started the winch – the machinery in the back of the lorry. We were all horrified as the horse's neck was stretched longer and longer. The noise of the winch was deafening and the horse was emitting involuntary groans. We were fascinated but repelled by what we were seeing. The body of the horse was eventually winched onto the lorry. We were all in tears when it was over and the crowd dispersed slowly. We never did find out why the horse died. We decided not to tell Mum what we had seen but as I was still crying when we arrived home she had to know. I had nightmares for some time afterwards. We were, understandably forbidden to go to the waste ground anymore and had to play in the garden or in the street outside the house.

About once a year a travelling preacher would arrive in the street and as the word went round we all gathered in the 'ring' (a circle of grass in the middle of one part of the road) to hear him. He usually had an accordion with him and we would join in with the hymns we knew. After struggling through two or three hymns, he'd then talk to us about God and Jesus for about quarter of an hour. Then he'd give us all a booklet to take home and move on to the other 'ring' in the road. I think that was the first time that I heard that Jesus had died to take away my sin and wanted me to trust Him.

There was usually half a dozen or so of us in our 'gang' and there would be crazes of games, which changed from week to week. We would chalk out a hopscotch court sometime if anyone had any chalk, sometimes we would find a chunk of it in the garden or someone would have a stick left over from their Christmas stocking. Another week it might be dibs or 'five-stones' if we could find enough of the right sort of stone. Our gang consisted of only girls so cat's cradle was always a favourite and I discovered recently that I still knew a lot of the patterns. We made dolls from worn out white ankle socks. The foot would become the head and body with the leg part cut off for arms and legs. We would cut up rags with which to stuff these creations and sew on wool – preferably yellow or brown – for hair. Mum usually had some scraps of material with which we could make clothes. These dolls were greatly loved, I think because we had made them ourselves. The faces we would embroider on with coloured cotton. We would give them names, of course, and built up quite a family of our own creation.

We liked Sundays because we usually had a roast dinner with roast potatoes; we didn't like sprouts or cabbage, though, but we always had to eat some of it. Often the 'Corona' lorry would come round and we would all go out to choose what drinks we wanted. It was usually Mum's choice but we just liked looking. Mum liked Cherryade but we liked Tizer – it sort of fizzed in your nose as you drank it. For sixpence a bottle it was good value.

2. High Days and Holidays

EVENTUALLY THE WAR WAS OVER, but it took a long time for things to get back to normal. We would go up to London to visit relatives and sometimes stay with them for a few days. One 'relative' was a lovely, quite elderly (or so we thought) lady known as 'Auntie Rice', who had been a friend of Mum's for some years. Twice a week she would load up a handcart with clothes that were kept in her back bedroom and push it round to where a street market was held. What a thrill it was to be allowed to go with her and help her display the goods as we thought they should look. There was a pickle factory at the bottom of this road, so we had the smell of cooking vinegar wafting to our nostrils. There would be other stalls along the road; we didn't like the smell of the fish van that sold jellied eels, whelks and cockles but we loved the tea van, especially when we were given a penny each to buy ourselves a 'cuppa' of very stewed tea.

We loved staying with Auntie. She lived in a long street called 'Meeting House Lane'; I always thought that it was called that because all the houses were joined on to each other, but found out that there used to be a Quaker meeting house there once. The lavatory was out in the back yard in a little shed – a magnificent throne like edifice with mahogany seat and newspaper squares dangling from a nail. It was very cold out there so we didn't hang around. The double bed that we slept in was so cosy with its big eiderdown and bolster type pillows; Auntie always reminded us that the chamber pot was under the bed and it was certainly better to use that than traipse outside in the middle of the night. Sometimes, as a treat, she would take us – usually one of my sisters, a granddaughter and me, 'up west'. We would catch the number 12 bus from Peckham Rye to Waterloo station then walk across Westminster Bridge. Years later when I discovered poetry, my sister Pam and I would always recite Wordsworth's 'Upon Westminster Bridge' while walking across it. My history lessons hadn't caught up with my poetry however and I didn't understand the 'ships, towers, domes, theatres and temples lie open until the fields and to the sky', where were they all? We would walk down the Mall to view Buckingham Palace; or feed the pigeons in Trafalgar Square. The capital was –and is– full of wonderful things to see and do.

We were always sent to Sunday School at the local Methodist Chapel in Ash where we lived. After the war ended, we especially liked the outings when two or three buses would turn up at eight o'clock on a summer Saturday morning. We clambered on board, squashed in three children to two seats; our paper carrier bags with our dinner in would be stashed away on the luggage racks; we would be counted more than once and then we were off. Bognor here we come! The noise that fifty children made was like a jackdaw colony but gradually the sound died down a bit. 'Let's sing' came a cry and we were off with 'One Man Went To Mow' 'Ten Green Bottles' and 'There Were Ten In The Bed' – all now politically incorrect but could we sing them! One of the adults would stand at the front doing the actions and we would all copy him.

Then suddenly we were at Bury hill where we had a short stop. The girls disappeared into one lot of bushes while the boys went the other way. We had a quick run around and then it was back into the coaches for the last fifteen miles. How we craned our necks to be the first to see the sea; the coach let us off by the pier at Bognor where it would pick us up seven hours later. We collected our lunches and macs – it might rain but never did- and followed the Sunday School Superintendent on to the beach. We made 'camps' with our friends and in no time at all were in our swim suits and running down to the water's edge. Some brave souls rushed straight in while others, I was among them, dipped a toe into the icy water. Eventually we would get in but apart from a few good swimmers we didn't take our feet off the bottom.

After an hour of this larking about we suggested that we had a sand castle competition. This was always great fun. We scoured the sands for artefacts to use, sea weed to adorn our finished castles, we hunted for shells and then started building. The tide was out so we had plenty of room, some boys were fed up so started burying each other in the sand; soon we had finished our construction work. There were many turreted edifices, some with tunnels right through, others with seaweed decorations, the tide soon turned and came in rapidly. The winners had sixpence to share between them and felt them-selves rich.

At four thirty we were all called together, counted – there was always one person who would have disappeared – and marched round to a church hall where we were going to have tea. Only bread and jam and a currant bun each plus squash for the children and tea for the adults but how good it tasted. Then there was a quick look round the fun fair, most of the children had spent their money but they enjoyed watching the horses going round or someone having a go at the rifle range.

All too soon it was six o'clock and again we were counted and shoe horned into the coaches. We were much too tired to sing and some folk even managed to doze off. We arrived back to be met by mums and dads who had been waiting for at least half an hour for their tired offspring. Within minutes the area was clear, the coaches had vanished, and within the hour we would all be asleep.

Dora, my eighty year old neighbour used to tell me how when *she* was a child, Sunday School outings were by traction engine and trailer usually to Savernake Forest. Boards, on which they sat were placed across the trailer, resting on the sides. One lady had taken her baby wrapped up in a white shawl; by the time she arrived home, the shawl was black with smuts from the chimney of the traction engine! They played rounders and cricket in the forest apart from exploring.

Another occasion that we looked forward to was the annual Christmas party at the chapel. An appeal for anything to put towards the tea was usually given, Mum gave some precious sugar I remember, and it was strange that in spite of rationing, we all took sugar in our tea. The afternoon would start with games, slap knee, postman's knock; hunt the thimble and another energetic game, one that I've never played anywhere else, ' I'm the umpsy bumpsy man,' then would come tea. Jam sandwiches were piled high on the plates and some of them were even strawberry, a real treat after the nondescript jam that was all that was available in the shops.

We soon cleared all the bread and jam and then the cakes were brought on. There was usually lots of slab fruit cake as it was sold by the pound in Woolworths, it was bright yellow with the fruit scattered rather sparsely through it. Rock cakes were very popular with butterfly cakes our favourite. Tea would finish most years with jelly and evaporated milk – a real treat. After tea we would play charades. As there were usually several adults helping they became team leaders and soon we would all be collapsing in laughter at the sketches. The evening would finish with 'Old Lang Syne', which we sang while standing in a circle and which necessitated a lot of stamping of feet.

3. School

PAM AND JOYCE WENT TO SCHOOL IN ASH but we lived in Ash Vale, which was about a mile away. Mum applied for me to go to Ash when it came time for me to start school but it was full; my name went on a waiting list and I was put down for Ash Common, which was closer but didn't have a good reputation. That too was full but they had started another school for the reception classes in the Victoria Hall, which was used for everything and about fifteen minutes' walk away. There was a lending library there and a 'slate club', which Mum went to every week to put sixpence on her slate card. This was drawn out at Christmas or whenever some extra money was needed. Sometimes a lorry would come during the day to sell 'Ticky Snack' steak and kidney pies. Mum declared that it was horsemeat but they made a welcome change to our diet.

The 'school' consisted of two classes one for the five year olds and the other was for the six year olds. There was just a big hall divided by a screen with a class on each side, which made concentrating very difficult, especially as the master who took the six-year-olds had a very strident voice. We had to recite our 'times tables' or else the alphabet, which always made a lot of noise. We left our outdoor clothes on a bench at the side. One morning it was raining hard and by the time we reached the hall, Mum and I were very wet. She decided to take my mac' home to dry and would bring it with her at lunchtime as I was only there for half a day. Unfortunately my hanky was in the pocket of my mac so I had to sniff.

"Eileen Collins, use your handkerchief!" called the teacher
"Please Miss I haven't got it."
"Come out here at once!"
"No! I don't want too," the tears started flowing.
The teacher stalked to the back row where I was sitting.
"We'll see what Mr Price thinks of that!"
I roared and screamed and Mr Price appeared.
"Whatever is happening here Miss Jones?"
"Eileen Collins has no handkerchief."
"Where is it girl?"
I had a job to stammer, "It's in my mac pocket and Mum took it home because it was wet."
"Do stop that noise, you'll come and stand out the front in my class for making such a fuss."
I roared louder than ever but was firmly led next door, where I had to stand at the front of the class facing the wall until break time.

The Victoria Hall was also used for a VE party for the children of the village. Everyone had contributed something towards the occasion to make it a memorable affair. We played games – to let off steam, it was a big hall so there was plenty of room to run around. There was a talent competition; I won sixpence for singing, "You are my sunshine," probably out of tune as singing wasn't my forte.

I was at that school for a couple of terms then Mum managed to get me in at the school where Pam and Joyce went. It was about half an hours walk but we could catch a bus part of the way, which if Mum had the money we could use. We actually went home for our dinner so we usually used our bus fare for one of those journeys. Going home after school was always an enjoyable walk unless I was sent to walk in front of Pam and

Joyce because they wanted to talk secrets with their friends. If it started to rain on the way home we would find a house with a porch and ask if we could shelter there until it stopped! It would take us a lot longer to get home than it did to get there. The school consisted of six classes, two were separated from the others by a partition, and these were the primary classes. We learnt the usual things plus knitting and sewing. I still have some of the mats that I 'embroidered' one's bright orange and the other brown, I suppose it was all that was available in the war. I knitted a purse in bright blue wool, just one long strip of garter stitch, which was folded and stitched up and I made a chain to go round my neck and stitched it on.

There was one teacher who we all used to dread. When she became angry her neck would go red followed by her face; we knew that this was the prelude to a telling off. If she went out of the room she would state, "There's to be no talking while I'm out of the room!" When she returned she'd ask, "Hands up all those who talked" *Hands would go up! We knew what would happen, she'd produce her cane from the desk drawer.* I nearly had it once but she let me off because I had only just returned from a bout of tonsillitis. She'd use it on one hand – the left one and everyone who had been talking lined up, licking their hands as it took away the sting of the cane.

We had to carry our gas masks with us in cardboard boxes that we wore over our shoulder. To encourage us to garden, there was about thirty small strips of land down one side of the playground, which the older boys and girls could use. I can't remember any of us having one but they were greatly sought after and tended during dinner hour by those who stayed to dinner. School dinners were introduced when I was about six; Mum said we could try them for a week but having been served bright orange semolina and lumpy tapioca, neither of which we had ever tasted, we decided we would go home for dinner.

The lavatories were at the bottom of the school yard and of the 'bucket' variety, so we didn't use them unless we were desperate. The girls had to run the gauntlet of the 5 to 7 year-old boys, whose open urinal we had to pass while they would be aiming to see whose wee would reach the highest up the wall!

We used to play games on the way home; we would allocate marks for people's front gardens and when the houses were near the road we would give marks for the front of them. Polished doorknockers were noticed and curtains, we would give extra points especially whether or not the steps had been whitened. We would explore odd things and once found some framed pictures in the bushes at the back of a dried up pond. We couldn't take them home or even tell Mum about them because we were not supposed to go near the pond, even though it was dried up. We often wondered what happened to them because the pictures were not there when we looked a week later.

We had the choice of ways to walk home, one way being longer than the other but much more interesting. There was a three-foot high brick wall running along outside one house and we could never resist climbing on top of it and walking with arms outstretched along the top of it. A bit further along was the blacksmiths forge, which drew us like a magnet. We would lean over the half door – I always had to be bunked up – and it was fascinating if a horse was being shod. We especially liked it when the blacksmith nailed a horse's shoe on. The smell was unique when he poured water over the foot to cool it down after the hot shoe had been nailed on.

Then there was the railway bridge. If we were lucky, a steam train would pass underneath us and we would inhale the sooty smelling steam.

Joyce stayed at that school until she was fourteen and then left to start work. Pam stayed until she was twelve and took an exam, which she passed, which enabled her to go to the Technical College in Guildford where she learnt Shorthand, Typing and Book-keeping, leaving there at sixteen. I left at eleven having passed the examination to go to the grammar school in Farnham.

4.   House Hunting

THE 'DURATION' EVENTUALLY CAME TO AN END and once again we were faced with the problem of house-hunting, as the original tenants of the house wanted to come back. Dad had some money when he was demobbed, a £50 gratuity, which sounded a huge amount at the time, but he found it difficult to find work. He wanted to be able to use his tailoring skills plus Mum's dressmaking ability and maybe set up on his own.

They found a property, which had an adjacent shop but it was only available to rent and there were sitting tenants on the first floor, an elderly couple who had been given notice to quit within six weeks. We lived there for a year and they were still in occupation when we moved on a year later.

The ground floor accommodation consisted of a front room, a living room and a scullery with the shop complete with a cellar. The garden was huge with apple trees and a fig tree, and a large barn. The toilet facilities were a bucket lavatory in an outhouse and no bath or washbasin at all. We washed in the scullery sink and had a tin bath in front of the fire once a week, taking it in turns to be first into the bath when the water was clean!

It was on the edge of the village and off the beaten track so no chance of setting up shop. We had to use the shop to store furniture as with only two rooms to furnish there was a lot that wouldn't fit in.

I don't know how the first floor tenants managed; they had to use our lavatory, it was very difficult living in such cramped circumstances.

One day walking home from school, Joyce suddenly shouted, "Eileen! There's something moving on your head!"

There was – and when Mum inspected all of us, she discovered that we had 'nits'. She went straight out and bought some Derbac soap, which was a grey colour and smelled like tar. Every day when we came home from school she'd have the kettle on and wash our hair. We hated it! Mum was far from gentle and it wasn't easy to work up a lather with such hard soap – lotions hadn't been invented then. We would then be sent to run around outside while our hair dried.

We would be dressed during the week in jumpers and skirts, to which Mum would attach a 'bodice' so we didn't have buttons or hooks and eyes to bother about. There were no zips in those days. The skirts were usually made from cast offs and of course being the youngest of the three of us, I always had what Joyce and Pam had grown out of. In the winter we wore plated lisle stockings held up by loops of elastic sewn on to our 'Liberty' (what a misnomer!) bodices and corresponding button on our stocking tops, most uncomfortable. If it was very cold Mum made us wear 'combs'- combinations, which were like a vest and knickers combined. We would still have to wear our vests and knickers, so using the loo wasn't easy. In the summer we would have cotton dresses made by Mum and grey socks for school and white ones for Sundays. when we would all be dressed alike with matching dresses and straw bonnets.

In the school holidays we would always have plenty to do but for a treat Mum would take us to Guildford for the day on the train. She hated travelling on the train but of course, we loved it. It wasn't a steam train but I can remember the excitement we felt while we waited impatiently for it to arrive. There was only one stop at Wanborough and the journey took about half an hour passing the unfinished Guildford cathedral on the way. It wasn't finished until after the war. When our son Peter graduated from Surrey University many years later, the graduation ceremony was held in the cathedral. The first stop when we reached Guildford was always the cattle market; not that we were interested in the cattle but there was always rabbits, chickens and ducks to see and the atmosphere was incredible. We would have a drink and doughnut from the refreshment van there and Mum would buy steak and kidney pies for our lunch later, which we would eat cold. We would then make tracks for the castle, although there was only the keep left standing.

The gardens were lovely and we ran around and then Mum let us climb up the spiral stairway to the top of the keep. There was a 'cage' built over the small standing area at the top so we were quite safe but we could see for miles and could pick out the cattle market easily and we waved to Mum, who was sitting by the goldfish pond. We ate our pies and cake followed by a weak drink of orange squash that Mum had tucked in her bag. Mum made us use the lavatories at the castle as we were going along the river next.

It was lovely there with cows grazing on the opposite banks; we could see water boatmen, beetle like creatures, swimming in the river. About a mile along the river there appeared what we liked best – an area of golden sand cascading down the steep bank into the river. It was quite shallow there, so safe to paddle; we would tuck our frocks in our knickers and although we had no spades we made do with sticks with which to dig. I suppose the sand must have been due to some geological fault. There was a café there so as a last treat Mum bought us a drink while she had a welcome cup of tea. Then it was back to town and the railway station and home. Sometimes if Mum had relatives staying, we would take them too and, as Londoners, they enjoyed the trip.

I became a Brownie. My Auntie Edna was the Brown Owl and I used to go to Brownies when I was four. I loved singing the songs that each 'six' had to sing. I was in the elves and our song was "This is what we do as elves, think of others not ourselves." The Sprite one was, "Here we come the sprightly sprites, Brave and helpful like the knights."

I went to Girl Guides when I was old enough and ended up as a patrol leader. I enjoyed the games we played and the things like 'stalking' and 'tracking'. I liked going to Church Parade and can feel even now the weight of the flag when it was my turn to carry the colours into church. The annual camp that was held on a farmers' field in a neighbouring village and although conditions were very primitive I thoroughly enjoyed it. We had to take three blankets, one of which was sewn into a bag and a groundsheet to sleep on. There were no seats, we cooked over open camp fires and ate our meals sitting on the ground. The 'lats' were just a deep trench with a heap of soil and a shovel; it was most difficult to use them and we all avoided it, choosing to 'go' behind a bush when we were out for a walk. Every morning there was the colour ceremony when the flag was hoisted on the flag pole where it flew all day to show that we were in residence; it was lowered at sundown after the camp fire. We would sit round the camp fire and sing songs – even 'Ging, gang, gooly!' One year I was lying down in the tent for the compulsory 'rest hour' after dinner, when I became aware of a tickling in my ear. I automatically poked my finger in it and realized it was an insect – it felt like a herd of elephants but eventually the earwig crawled out. My sister Pam became a Sea Ranger; they had their headquarters along the tow path of the Basingstoke canal and used to go rowing a lot as well as learning how to tie knots.

Dad managed to find a job in the tailors shop in one of the barracks in Aldershot. Every few years, the army contracts were renewed for the tailor and might be given to someone else other than the man that was already in residence. Sometimes Dad would be able to stay on and work for the new boss, if not he had to start job hunting. He was able to do dry cleaning so would do that for a couple of years until a tailoring vacancy occurred. Mum did out- working for another retail army tailor in the town; making beret bows was one job she had to do. They were tiny things and served no useful purpose but army berets had to have them. Mum also had a couple of house cleaning jobs as well as taking in sewing. Joyce left school and obtained a job in a local grocer's shop where she stayed until she was married some twenty years later. Pam won a scholarship to go to Guildford Technical College where she did a secretarial course and eventually obtained a job at the Royal Aircraft Establishment at Farnborough.

Owing to the overcrowding, we were allocated a council house that had been used to house Italian prisoners of war; this was in Tongham, another small village. The house was semidetached with three bedrooms and two rooms and a bathroom downstairs. It was lovely to have space at last. Dad dug up the garden, which was quite large and he planted flowers in the front garden and vegetables in the back one. The one trouble with the house though was cockroaches! They would come out at night from under the skirting boards and if we had to come down at all they would scurry away with a loud rustling sound. Mum bought some cockroach killer, and kept her flat iron handy to kill any that were still moving in the morning; often there would be a trail of powder leading under the piano where the cockroaches had walked.

We always had a cat at home; Mum said that it would keep the mice away. We never had mice so I suppose that it worked. They were always tom cats but Mum never had them neutered as it cost too much money. The current cat was quite old now and didn't seem well so Mum decided that it was time to get rid of him. There was no thought of taking it to the vet; there was a slaughterhouse about quarter of an hour walk so the three of us – I was about eight at the time – set off with Tommy in a box. We could smell the place before we reached it. We walked up a road leading to the big doors and heaped up at the side were piles of bones that looked as if they would come from an elephant! We found a door that read 'Office' and Joyce knocked on it timidly. It was flung open.

"Yes?"

"Please Mister, Mum says could you put our cat to sleep?"

"Yes! Have you bought a shilling with you?"

"Yes sir, here it is."

Joyce handed over the money and the box with the cat, who by now was yowling loudly. We were glad to get home but felt very sad that Tommy had had such an ignominious death.

I made friends with other girls living on the estate and discovered that one of them went to a dancing class on a Saturday afternoon. This sounded like fun so Mum said that I could go too. The lady who ran it charged sixpence each and as I had a shilling for pocket money, I paid the money myself. I had had visions of flitting around in a tutu – in spite of being tubby and decidedly overweight. Although sometimes the lady in charge gave us a demonstration of ballet dancing, it was actually ballroom dancing that she taught. Still, it was fun and I kept going for some weeks. My sister Joyce went to dances sometimes at a hall connected to the local Catholic Church and occasionally they had a 'social' instead of a dance, which Mum let Pam and I go to. The difference between a dance and a social was that games were played in a social interspersed with dancing; there was also the ubiquitous raffle but we could never afford a ticket. We danced with each other as the men danced with the women. All too soon it would be time for the last waltz and the National Anthem, for which we all stood and sang lustily.

I was invited by a girl along the road with whom I had become friendly, to attend a Bible Class at a small Evangelical Church in the village. There were quite a lot of children in the Bible Class and Sunday School but it was run on very traditional lines, the lady who taught my class was probably in her forties and always wore a hat. One week she asked me if I'd like to go hop picking with her. There was a farm in the next village that grew hops and I even managed to get up early to cycle the couple of miles to Runfold where the hop fields were. In the early morning it was a magical place to be and I'd never before been anywhere with such an atmosphere. The grey dawn light was sloping through the rows of hop bines, making the field look mysterious. People were dragging big square baskets into the alleyways plus prams with not only a baby but loaded with clothing and bags plus the odd tea pot and kettle. At 8a.m. a whistle blew and immediately everyone swung into action pulling down the 'bines' that slanted across each alleyway. Each family was allocated an alleyway. Most of the grownups stood round their big basket and picked the oval shaped green hops straight into it. The children would be given smaller containers, which they picked into and emptied into the big basket at intervals. The little ones started to charge up and down the alleyways playing tag or hide and seek until they got in the way. As the morning went on the smell from the hops was overpowering. One person from the various groups was sent to get a kettle of water, which was placed on a little fire that had been made where the bines had already been picked. Nobody else stopped working but were all picking at such a speed that their fingers were a blur. They gradually worked their way along 'their' alleyway dragging their containers with them. The children collected 'hop dogs' – huge green caterpillars and made them race against each other. The pickers would come to a break in the bines but continued in the same alleyway the other side of the track. I wasn't any good at picking and never repeated the exercise.

As the afternoon wore on, the pickers surreptitiously pulled extra bines down so when the call rang out "No more bines" they quickly pulled one last one and still had an hours work to do. When the last hops were picked the ladies would reach down into the hops and fluff them up, thus making it look as if there were more hops in the basket. A tractor and trailer came round taking stock of what each family had picked and the tally entered into a book, the tallyman was not a popular man I'm afraid. The smell of hops is indescribable, pungent with a strong 'beery' scent. Hands were stained dark green, which was hard to remove. There were no modern facilities there, hands were wiped down the sides of trousers and further up the alleyway had to suffice for a lavatory.

5.   Social Club

MY SISTER PAM AND I BEGAN to go to a weeknight club at the church. A couple with a young baby moved to Tongham and also attended the church. Joan began a girl's Bible study class during the week, which we went to in her flat, and there I realized that I was a sinner and needed a Saviour and trusted the Lord Jesus as my Saviour. There was quite a group of teenage girls and once a month we could go with one of the leaders to a Youth Rally at Aldershot Park Hall, which was a much bigger church. We would all pile in Doug's (one of the leaders) old van, about half a dozen of us. These services attracted about a hundred, mainly young, people and we nearly raised the roof with our singing. The proceedings were conducted by a man, Mr Sullivan, whose nickname was 'Pop' he believed what he was telling us and had a terrific sense of humour. He was 'interviewed' one evening and told that if he didn't want to answer a question to say, 'Steamship'. The first question was "what is your name?" The reply came like a flash, "Steamship Arthur Sullivan!" – His first name was Nathaniel! I didn't have a clue then that he would be my father in law!

At last we were able to put down roots; Dad was in work and Mum was still able to make her beret bows and did dressmaking. She also found a cleaning job nearer to where we were living so we felt quite rich. We lived in that house for about four years. When I was eleven, I won a scholarship to a Grammar School in Farnham and found it tough going. It had been a private grammar school and some pupils were still fee paying, which made a rather 'them' and 'us' situation. We had to catch two buses to get there, which took about an hour then we had to walk for quarter of an hour once we arrived at Farnham. The uniform was changed frequently from box pleated gymslips to a pinafore style one. The blouses were white square necked ones, which were changed to blue aertex shirts; Mum couldn't afford them at first so I still had to wear the white ones until I grew out of them. We had to wear navy blue knickers – bloomers would be a better word – which were used for physical training with the white shirts. The trouble was that with constant washing (I only possessed two pairs), they ended up grey. We had sewing classes; the first thing we made was an apron for cookery, red and white checked gingham. The next thing was a pair of knickers from lemon coloured cotton with no 'give' to it. We were told to work in pairs and to measure each other's hip size, which created a great deal of merriment – and embarrassment- most girls were 30" or under. I was 36"! I decided I couldn't be that big so said that I was 34." Needless to say they never fitted and Mum was cross because of the waste of money.

Most girls had their own tennis racquets and after a lot of persuasion, I managed to persuade Mum to buy one for ten shillings from an advertisement in the paper, I never had much chance to use it, as tennis, like the other sports we played, were just not my 'thing'. I didn't like hockey and was hopeless at it but I still had to play. I joined the Dramatic Society, which I enjoyed and played a rabbit in the production of 'Toad of Toad Hall' with a pair of rabbit ears, shorts, jersey and tie. The next year it was 'Alice in Wonderland' and I was a 'Playing Card.' I had a big card pinned to me front and back, my white school blouse and white knickers, which belonged to somebody else as I didn't possess any, only navy blue ones! As the school was so awkward to get to, nobody from home ever came to anything that was on at school.

I still have a friend dating back to my time at Farnham and we remain in touch. I discovered that she too attended Aldershot Park Hall, on Sundays as there was a Bible Class held in the afternoon. I started going with her after talking with Joan at church about it, she was in favour of it as it would put me in touch with more people. There were about thirty teenagers in the Bible Class, which was affiliated to the Covenanter movement and very lively. 'Pop' Sullivan was one of the joint leaders with 'Jairus' being another; the girls' leader was 'Hig.'

It was nice going to something that was specially arranged for teenagers. We began the afternoon singing 'choruses' and sang with enthusiasm taking the various parts when there was a song that needed it. Someone would have been asked to give a 'testimony', which was telling how they had trusted in the Lord for salvation. Another afternoon there would be a quiz, boys versus girls, which was always great fun. The last twenty minutes we divided into three classes with the different ages and boys and girls.

On Bank Holidays a hike was always arranged by the young people, about twenty of us would meet up outside the Gospel Hall and we would start walking. Frensham Pond was our favourite place to go, a distance of about ten miles; the first couple of miles were still in a built up area. After walking for a couple of hours we would have a short break for a drink then reach Frensham by dinnertime. We would eat our packed lunch and if it was hot have a siesta or paddle in the pond, but soon we were playing an energetic game of rounders or 'crocker', a game played with a baseball bat and rugby ball. I'm not sure if it's still played by church groups but it was THE game to play in the fifties. Then we would just sit and talk or go for a walk before finishing our food and starting on the way home. One year it came on to rain and although cars were not plentiful, any that came along and going in the right direction were waved down and usually lifts obtained, two at a time for at least part of the way back to Aldershot.

Mum and Dad had decided that it was about time that they became mobile so bought a tandem! Mum had never been on a bike before but the tandem was ideal for her as the handlebars at the back were in a fixed position. They even put a seat on the back for me until I grew too big for it. When I was older and had been out for the day on an outing, Dad would ride the tandem to wherever I was arriving and I'd sit in Mum's place. I found it quite difficult as the chain was connected to both sets of pedals so if legs became tired, there was no slacking. Also as the handlebars were fixed it was hard not to be able to steer it. One day when I was about twelve I had gone on a coach trip to Southsea with a school friend whose parents belonged to an Old Time dancing club. We had a lovely day by the sea but on the way home, stopped at a pub near Petworth where the function room had been hired for a dance. By the time we left there it was ten o'clock and Mum was expecting me about eight and I was going to have to walk home. I was so pleased when we arrived in Aldershot to see Dad and the tandem. When I grew out of it they decided to have a motor fixed on the back; this gave them a lot more scope and they toured the South of England on it, for holidays. They even rode it in London when we moved some years later but of course there wasn't the traffic that there is now.

Mum and Dad wanted to move again; they found another family from an estate back in Ash who wanted to move to

Tongham so we exchanged houses and moved into Long Acre. I was thrilled to discover that Anne, my friend from school lived next door with her parents plus two sisters and a brother. This was a much larger house and garden; the estate wasn't yet finished although the roads were all in place. Anne and I both had our tennis racquets and spent hours in the holidays playing 'tennis' on the empty stretch of road. We did a lot together, now that we were twelve we both had bikes and as long as we were home by a certain time our mums didn't mind us going off for bike rides. With some other friends we would go blackberrying; Mum didn't like cooking blackberries so I'd take them to Hig our Bible Class Leader.

6.　Leaving School

I DECIDED TO LEAVE SCHOOL WHEN I WAS FIFTEEN. Anne was going to leave as well although she wouldn't be fifteen until after school had broken up for the Easter holidays. Money was always tight in our house and I decided that it was time that I contributed to the family purse. I managed to get a job in the office of a hospital laundry about half an hour's cycle ride away. There were two other girls working in the office and from what I gathered had complained that they had too much work to do, which is where I came in. Unfortunately, I wasn't really needed and spoilt their set up. I became the butt of snide remarks and had a total lack of communication with them; after three months I could stand it no longer so decided to leave. Mum, of course was cross but a big Timothy Whites and Taylors warehouse had opened on the outskirts of Aldershot and I managed to get a job as a stock control clerk. This consisted of finding items, which were listed on cards, on a shelf and counting the items, entering the quantities on separate cards, which went to a checker to work out if more stock was needed. It was nice working with people who didn't resent my presence and as we had a tea break we were able to get to know each other. I worked in the 'patents' department, which consisted of drugs, ointments and other medications. I had an immediate superior who was an older woman but easy to get on with.

One practice that took place was ear piercing! This took place in the Ladies toilets at lunch time, which was an hour long, I went home for lunch. The implements for the ear piercing were a darning needle and a cork. Girls could sometimes be seen in the afternoons with bright red ears and maybe a trickle of blood. Ear piercing as it is now was unheard of and the only sort of places that did it were men's barber shops or a 'tattoo artist' of which there were many in Aldershot.

As I was growing up, so was David, but being five years older than him, he was just one of the boys although I'd often be asked to his house for tea on Sunday by his father, as it saved me going back to Ash and back again for the evening service. Dave always seemed to be doing something; he made a really good go-cart and as some of the streets near where he lived were quite steep was able to put it to good use.

When David's father had been demobbed from the army, he bought a grocers shop in Aldershot. It was exactly like 'Ark-wright's' in 'Open all Hours' and Dave remembers how his Dad used to bring things from the shop in the morning and prop them outside. The family lived over and behind the shop; it was quite small with a living room and scullery at the back and three bedrooms upstairs. There was no bathroom and the flush toilet was outside the back door together with a shed for storage and a small garden. The tin bath hung on a nail in the shed and had to be dragged into the scullery to use. There was an electric copper in the shed from where hot water had to be fetched, poured into the bath and topped up with cold from the tap. The bath water didn't stay hot for very long and the secret was to have it as hot as was bearable. The three children had to share bath water with Dave being usually last in the pecking order. 'Pop', as he was known universally, decided to turn the shed into a 'proper' bathroom with a bath that could be emptied via a plug hole. He managed to buy a 'proper' bath at a second hand shop and made a soakaway in the garden with a pipe going into it from the bath. Cold water now had to be fetched from the scullery but it was a lot better than before but at least there was hot water on tap to top up the bath..

The shop sold almost everything. Women would come in, order book in their hands and would leave them with Pop returning later when the order would be packed in a box ready for them to take home. They usually paid weekly after their husbands had been paid as often the ladies would need something apart from their main order.

"What do you want today Mrs Jones?"

"I've run out of potatoes Mr Sullivan. Could I have three pounds please?"

"Here you are; hold your pinafore out."

The potatoes in the scale pan would be tipped into the out-stretched pinafore. Another satisfied customer.

When Dave was old enough his father bought a trade bike so that orders could be delivered. The shop was at the bottom of a very steep hill so often he'd have to push it up to the top but be able to coast down. It was nice to have a job and as the customers gave him a few coppers for delivering the goods he had money in his pocket. The shop, unfortunately, didn't provide an income sufficient to feed two adults and three children so Pop decided to get a job and pay a girl to run the business. This arrangement worked quite well as one of the Bible class girls needed a job and Joan seemed just the right person. Eventually the business was sold and the family moved to a three bed-roomed house and Pop found a job as a Civil Servant in the army camp. The girls changed schools as the house was in a different part of Aldershot, but Dave had won a Scholarship to go to the Grammar School in Farnborough, so for the summer term he stayed at the same school.

Dave quickly made new friends in the road and found ways of supplementing his pocket money mainly by collecting horse droppings and selling the manure by the bucket full to the keen gardeners in the road. One of his friends was Bobby Isaacs who lived a few houses along the road from Dave.

7.   Hunting Jackdaws

IT WAS THE SCHOOL HOLIDAYS. David and Bobby were bored.

"I wish we could get hold of a parrot. I'd love to teach one to talk." Bobby said as they sat on the kerb at the side of the road.

"Don't know where we would get one from, Bob," replied Dave, "I've heard that if you get them young enough, you can teach jackdaws to talk. Why don't we try and catch one?"

"Great! Do you know where there are any?"

"I know! Waverley Abbey. There's bound to be some over there."

Waverley Abbey was about three miles from Aldershot so they fetched their bikes, took some water and a couple of apples, and set off.

There was a chestnut paling fence all-round the abbey, which was just a ruin and yes, it was inhabited by jackdaws. They lifted their bikes over the fence and clambered over themselves. They surveyed the situation.

"Look Bob, the nests seem to be all just below the top of that wall. We could climb up easily I'm sure," ventured Dave, "I'll go up and see if there's any eggs."

He started gingerly up the wall but ran out of handholds.

"It's no good Bob we'll have to find another way up."

"Let's go round the back of the wall, it might be easier from there. We could climb down from the top of the wall," replied Bob.

They walked round to the other side conscious all the time of the birds circling around above their heads.

The boys started climbing. This was better than the other side; Dave reached the top and peered over the edge followed by Bob. Yes, there was a nest just below them but they couldn't see how to get to it.

"I'll try Bob, it looks as if there are two young chicks in the nest. Here goes." Dave swung his legs over the edge and felt around with his foot. "There doesn't seem anywhere to put my feet. You have a try."

"You're chicken." jeered Bob, "I'll show you how to do it."

Dave climbed back up gingerly, feeling rather dejected.

"See if you can do any better."

Bob started to climb down. He was bigger than David and considerably heavier and found it hard going.

"Help! I'm falling!" he hit the ground with a thud.

"You O.K. Bob?"

Bob groaned. "My arm hurts Dave."

David clambered down as fast as he could. Bob's arm looked funny and there was a stone sticking in it.

"Bob, I'll have to go for help. Don't move will you."

Dave took off his jacket and covered Bob up; he had heard somewhere that you had to do that.

"Don't be long Dave," groaned Bob.

He climbed over the fence and remembered that they had passed a cottage on the way along the lane. He ran as fast as he could and banged on the door.

"Please Missus, my mate's fallen off the abbey wall and hurt his arm. Have you got a 'phone and could you get an ambulance please?"

"You're not supposed to go in the abbey grounds," replied the startled lady.

"I know we shouldn't have done but we wanted a jackdaw. Bob's hurt bad. Will the ambulance be long?"

"I'll ring them now. You'd better wait by the abbey fence so that you can guide the ambulance men in."

"Thanks Missus."

Dave ran back to the abbey, climbed over the fence and went to where Bob lay. He seemed asleep and was very white.

When David heard the ambulance coming he went out in the lane to meet it.

"Where's your mate then sonny?"

"You'll need a stretcher and it'll be a bit hard getting it over the fence."

Dave led the way back to the fence. The ambulance men managed to lift the empty stretcher over the fence.

They examined Bob carefully and pulled his arm straight. The 'stone' went into his arm.

"What about that stone in his arm?" asked Dave.

"That's a bone; it'll have to be dealt with properly at the hospital."

The ambulance men strapped Bob on to the stretcher and covered him up with a blanket. They had a job getting the stretcher over the fence; Bob was conscious now and looked very white.

"See you soon Bob."

"We never did get that jackdaw Dave," croaked Bob.

The ambulance men closed the doors.

"You'll tell his Mum won't you sonny?"

Dave gulped! Bobby's Mum! He had forgotten about her. He was quite scared of her as she shouted sometimes.

He collected both bikes and managed to ride his, holding Bob's with one hand while he rode his own.

He walked up Bob's path with Bob's bike and knocked on the door. It was flung open.

"Well?" queried Bob's Mum.

"Please Mrs Isaacs, Bobby's had an accident and broken his arm- it's all right though, the ambulance is taking him to Guildford Hospital!"

To Dave's dismay, Bob's big frightening Mum burst into tears.

"Come in David and tell me what happened," she sobbed.

Dave told her all about the accident and gradually her sobs subsided. She went to a cupboard and pulled out a bar of chocolate.

"Here you are Dave, you deserve it."

She seemed totally different after that and when Bob came out of hospital, Dave often went round to keep him company. Bob had his arm in plaster for months, which all his mates signed, and every so often he had to have an operation as he had broken two bones and one bone was growing at a different rate to the other one. Dave can never see a jackdaw without being reminded of that day.

8.   Coronation Year, 1952

LEADING OFF THE ROAD WHERE DAVID LIVED was a brickworks. There was a pond in the premises and water had always attracted boys of whatever age and Dave and his friends were no exceptions. In the summer when the men had finished work, they would climb over the fence and make for the pond. There was usually enough boys to divide into two teams and having armed themselves with the necessary equipment they would take the 'whippiest' stick they could find, make a ball of mud to go on the end take up a position on the edge of the pond and battle would commence. The sticks would be pulled back then released, sending the mud ball flying across the pond. Dave's mother was never very happy when he'd arrived home plastered in mud.

Dave had another friend named Geoff. They both had bikes, Dave's parents bought David's when he passed the scholarship to go to the Farnborough Boys Grammar School.

They rode them everywhere. They rode across London once when they went to a boy Covenanter camp, which was being held in Norfolk. Whenever there was a Sunday School outing to the sea they would pedal their way down to Littlehampton or wherever it was, a distance of some thirty five miles. There would usually be a stop at Bury Hill and they would catch us up there and then would beat the coach to the sea.

Everyone had bikes then and used them. At one time I was riding eight miles to work no matter what the weather was like. I started off with a second hand bike then managed to save thirty pounds for a new one, which I really treasured. I rode to church as well, even in a skirt. When I started earning money I bought a lovely pink organdie frock, which was lined with pink taffeta and I even cycled in it to church on Sundays completing the ensemble with a pale pink hat to match and white net gloves and sandals. It was totally unsuitable for riding in but I loved it so much as most of my dresses were made by Mum.

It was 1952 and a most exciting year as the Queen was going to be crowned. Pam was staying in London with her friend the night before and my friend Anne and I were going on the early morning train to be there. The train was packed but it was a very happy carriage load of people even though we were packed in like sardines in a can. We all tumbled out when we reached Waterloo Station and joined the thousands streaming across Blackfriars Bridge.

We had decided to go to Trafalgar Square as the procession would surely be going slowly there to negotiate the corners. It seemed when we arrived there that everyone else had had the same idea as there was just a sea of people. We found a toilet as we didn't know when we would have the chance again. We then wriggled our way into the crowd finishing up about fifteen rows from the front. We could only see the backs of the people in front of us but it didn't matter, we were there and the atmosphere was terrific.

The radio programmes were being broadcast through loudspeakers and as we listened we heard the news that Edmund Hillary and Sherpa Tensing had reached the summit of Mount Everest. A huge cheer went up from the crowd and although we hadn't even known the attempt was being made we realized what a terrific accomplishment it was. The hours ticked by, we could hear various carriages going by and at last we heard that the procession had left Buckingham Palace. As it approached we could hear the roar of the crowds cheering. We craned our necks and could just see the top of the coach as it went by. We were deafened by the cheers of the crowd but we were shouting as loud as anybody. Soon all the processions had passed. Anne and I decided to go home as my parents had recently acquired a television set. We caught a train straight away and an hour later we were home and the Queen was still in Westminster Abbey. We settled down with Mum and Dad and Joyce and enjoyed watching the rest of the day on the black and white set with a twelve inch screen. Pam and her friend had spent the night opposite Westminster Abbey and had managed to see quite a lot of the comings and goings during the day.

I became the Leader of the Jucos (Junior Covenanters) and was given the nickname of Berry because I had the habit of wearing a black beret! It was thought more respectful to use a nickname than a Christian name and not as formal as using Miss Collins. The Bible class and the top classes of the Sunday School on Sunday afternoons were affiliated to the Boy and Girl Covenanter movements. We often went up to the annual rallies, which were usually held in Westminster Central Hall in London. We would catch the 'workmen's' train, which left Aldershot at 6.45.a.m. As I lived outside Aldershot I spent the night with another friend, Mavis, as she lived twenty minutes' walk from the station. Sometimes, another friend, Hilda would join us as she lived at Farnborough. We would all three cram into Mavis's double bed and didn't get much sleep as you can imagine. I used to make little pin curls in my hair and to keep the hair grips in would wear my black beret in bed! I didn't bother to take my beret off and do my hair until we were in the train and then Mavis enjoyed doing it for me.

We wouldn't bother with breakfast and after a quick cup of tea and a wash would set out in the grey light of an early dawn to get to the station in time. We would meet the other girls and there'd be about a dozen or more of us. The first place we went to was Lyons Tea Shop at Westminster; I always had a poached egg on toast as it was the only place I ever had one. Then, having decided on the train which area we were going to visit, would set off on our sightseeing tour. If there were any girls who hadn't been to London before we would work out a route taking in as many of the sights as possible and decide how we could get the most out of the four hours or so that there was before the rally in the afternoon.

We would walk up Whitehall, stopping off at Downing Street, which in those days you could walk along and take photographs of No. 10. Then we would carry on along Whitehall pausing for more photographs with the guard on his horse at the entrance to Horseguards Parade. Our next stop was Trafalgar Square, cameras at the ready for photos of the girls with Landseer's lions and, of course the pigeons! They were so tame that for a handful of bird seed, would even perch on the girls hands. Then we were off to look at Eros in the middle of Piccadilly Circus. We caught the Underground train on which some girls had never travelled, to St. James's Park for lunch; we had all brought sandwiches with us plus a bottle of drink.

We soon ate our lunch so just had time to look at the ducks on the lake and visit the toilets before finding our way to Westminster Central Hall for the Rally. The building would be packed with girls from all over the country and it was nice to meet up with girls we had met at camp. The singing was wonderful. Dave and his friend Geoff cycled there one year and arrived after the service had started: they were both wearing shorts and looked hot and bothered. They sat up in the balcony and had to edge along the row to their seats, I was most annoyed with Dave for turning up at something which was a girl's domain, especially as he was wearing a bright green jersey and shorts, although there were other males in the meeting. Various girls or their leaders would give reports of the camps that had been held in the summer and any special services that had taken place. There would be a good speaker who would read from the Bible and tell us more about God's love for us and that we needed Him in our lives. We had a 'cuppa' from a stall outside and were soon on the train again. By the time we reached Aldershot we would be very tired but would have had a lovely day. Some years we would enter the Girl Covenanter netball tournament, which was usually held at Clapham Common we were runners up one year. We practised for this every Saturday afternoon when we could have the use of the netball courts at the local High School. We all enjoyed this and liked each other's company.

In 1951 a group of us went up to the Festival of Britain on the South Bank of the river Thames. All that remains of the site now is the Festival Hall but then there was the Dome of Discovery with all the different things that had been invented or discovered down through the years, the Skylon a weird column reaching up towards the sky, and other very modernistic things to see. We went over to Westminster Abbey, which we enjoyed seeing; two of the boys went back inside after we had left it, as they were arguing as to whether the Stone of Scone was still under the Coronation Chair. It had been stolen and taken back to Scotland so the boys wanted to make sure that it was back where it belonged – under the Coronation Chair.

9. Summer Camp

EVERY AUGUST WE WOULD ARRANGE A CAMP for the Covenanters and 'Jucos' (Junior Covenanters). We used to go to camps organized by The Girl Covenanters Headquarters. Four of us fifteen year olds went one year to one that was being held near Hitchin in Hertfordshire. This was quite a big adventure as we would have to get across London from Waterloo to King's Cross and find our way from the station when we arrived at Hitchin. I remember that it was a very hot day and we were glad that we had been able to send our luggage in advance. This was a useful service and for a small charge luggage could be taken to the station at Aldershot and sent on to be delivered to our destination.

We met up with other groups who had travelled on the same train, and discovered that there was a two mile hike to get to the camp site but we were glad to stretch our legs and with a bit of singing we were soon there. The camp was based in a big barn where we would eat and have the main activities; there were about six or seven bell tents in an adjacent field and then there were the cow sheds! They were devoid of cows and had been well washed out but there was still an aroma of the previous occupants, this was where some of us were to sleep.

We had all been asked to take a jar of jam with us, which Mum thought very strange. The idea was that there'd be a lot of home-made jam, rationing hadn't long finished and the variety would be nice. We were divided into teams of about six or seven and after finding our luggage had to fill our palliasses with straw, Mum had made my palliasse from a couple of old curtains sewn together others had brought mattress covers, no sleeping bags around then. It's not easy to get the straw in the right place and of the right density and soon we were smothered in it. The ground was very hard and we were glad of the straw underneath us. We were placed into various teams all named after insects beginning with the first few letters in the alphabet – Ants, Beetles, Cockroaches, Daddy Longlegs, Earwigs, Fleas, I was a Daddy Longlegs and was sleeping in the cowshed. Hilda and I were together and the others were in the Earwigs.

We had to make up our beds and then it was time for a meal, which was eaten in the big barn. We started by singing grace, "Blessings and Glory, Thanks and Praise, Offer to God the Giver, He is our strength for all our days, He is our Joy for ever." Later in the evening it would be time for Campfire, which would be lit well away from the tents. We would get into 'evening dress' for this – pyjamas with one of our blankets wrapped round us. The sound of about fifty girls singing in the open air was wonderful. After the singing there would be a 'testimony' from someone telling how they came to trust in the Lord Jesus Christ and what He meant to them. To finish with, 'Chief' who led the camp and devotions gave a challenging talk on what it really meant to be a Christian. We all went thoughtfully to bed.

During the week we went swimming, for hikes, treasure hunts, shopping in Hitchen but underlying the whole week was the fact that Jesus loved us and wanted us to love and follow Him. The week passed all too quickly and we were soon on the train back to London then the Underground to Waterloo and home.

The next year three of us took the "Jucos" (Junior Covenanters}, away to a camp in Great Holland in Suffolk, which was held in a church hall. It was a long way to go but we hired a van, which had nothing to sit on but the floor! It was a very long way and we had to travel through London, no M25 in those days, which, as it was a very hot day, was very trying. Some of the children dozed off, which was as well as it took us about five hours to get there. When we reached Great Holland we met up with other groups of girls, about fifty children in all with ten helpers. The man who drove our transport had a cup of tea then set off to travel back to Aldershot.

We enjoyed our time away and it was nice to get to know the girls in a different environment. We were a half hours walk from the sea at Frinton so spent part of most days on the beach. We had a coach trip one day to Felixstowe, which was about fifteen miles away but it was very busy and we preferred the beach at Frinton. The journey home didn't seem quite so bad; most of the Jucos slept part of the way.

The next year we decided to go to camp but realized that it wasn't really practical to travel so far. We asked around and found a Christian organization The European Christian Mission, that ran camps for children. We investigated this and for the next three years we went to their camps. This was a camp that slept in tents, six to a tent and the first one we went to was held at Sherbourne St. John near Basingstoke.

Sleeping at the same camp but working for the farmer who owned the field was David. His father had asked Mr Hatt, the farmer if David could spend the summer holiday really seeing what farming was like! He asked Mr Hatt to give him the worst jobs so it would put him off farming. Mr Hatt recalled this years later and said that he gave Dave the dirtiest and most tedious jobs he could find, Dave loved it all. It was convenient for him to stay at camp and he could cycle the three miles to the farm each day.

It was a happy camp and the twelve girls we had taken really enjoyed it. We travelled the twenty miles each way on the back of an open cleaned out coal lorry! We couldn't decide how to get there and Mum had the bright idea of asking the coal man. It had been well swept and as we clambered aboard with all our luggage we were glad that it wasn't raining. The modern Health and Safety rules would make it impossible now, as we broke all the rules, the sides were only about three feet high and we just sat on our bed rolls on the floor, it's a good thing it wasn't raining.

There were about thirty other girls on camp plus a dozen or so helpers. We had kit inspection every morning when the girls lined up outside their tents with their crockery, folded blankets – and Bibles on display. We played games had a couple of trips into Basingstoke and went to see the ruins of a manor house that had been burnt to the ground in the civil war. (Why we went there I can't imagine).Once again we had the customary camp fire, singing and Bible talk.

The next two summers we went to Arundel and Bognor with the same people. The Arundel site was in a field at the junction of two rivers. One night there was a storm and we all had to get dressed and help to hang on to guy ropes to stop the marquee from blowing away. The children's clothes all got damp and we had the job of trying to dry them out, I've never forgotten the experience and often wonder if the girls remember.

It was at the Bognor camp that Brenda, my colleague, and I decided to try to organize our own camps. We thought that we would have to utilize church halls as tents would be too hard to obtain and manage. I wrote to various churches in different sea side towns. We had a few replies but a Congregational Church in Folkestone had facilities that looked ideal. It was too far for us to go to have a look at the place; my friend Hilda now lived that way so she visited it and felt that it would be ideal for our use, it was. This time we travelled in style and went by train having sent our luggage in advance by rail a few days before.

Saturday morning everyone congregated at the station, each with their lunch as it would take about three hours to get to Folkestone. We caught the train to Waterloo and then had to transfer all twenty five girls, which included Covenanters ("Covies") as well as Jucos to Waterloo East. There were about eight helpers including our two stalwart cooks who kept us well fed on very little. It was certainly a help being in the middle of the town. We had arranged with the corn chandler whose shop was nearly next door, to have a quantity of straw available to stuff into our palliasses. Nobody suffered from hay fever thank goodness and soon beds were made and we took everybody down to see the sea.

When we arrived back at the Church Hall we were able to sit down to baked beans on toast and cake. Several of the mums had made cakes for us and they had travelled quite well. Every day I would write something called 'Camp Diary', which was an account of what we did during the day – written in rhyme! We had the first instalment before the Jucos went to bed but it was a long time before they went to sleep. The Covies went to bed a bit later but we were all tired after our long day.

We did different things each day. We had P.E. first thing, which was not to be taken seriously and after breakfast we had a 'Quiet Time' when we would get the girls together in groups and read the Bible, talk about what we had read and have a prayer. The youngsters had the choice of shopping, roller skating, swimming or the play park in the morning In the afternoon we all went to the beach and had more organized things for them to do. One day we took them on the Romney Hythe and Dymchurch Miniature Railway, which runs from Hythe to Dymchurch across very bleak countryside – just stones with the occasional shack. We had our packed lunch at Dymchurch where there were two lighthouses as the sea is receding there and the original one was a quarter of a mile inland. The wind always blows there as well. We were quite tired when we got back to Hythe so we caught a bus from Hythe to Folkestone and walked the rest of the way, once we had had our tea we felt fine again. One evening we would have 'Camp Concert', which was always fun, when the girls would work out a sketch, song or recitation to perform. Another evening we would have a 'midnight feast', not at midnight but quite late when they would eat all sorts of highly indigestible food they had bought. Another night some of us would take the 'Covies' for a walk along the prom in the dark and we felt we were really getting to know these teenagers. Some of them over the years became leaders themselves and It never ceases to amaze me that so many of those girls are still involved with a church. Peter, our younger son lives in Aldershot and attends the church where we grew up and were married; many of our contemporaries are still there.

We went to Ryde on the Isle of Wight the following year and stayed at a large house that was owned by a local church. It was very exciting having the trip on the ferry after a train journey to get there and we soon settled down into the various groups. The Covenanters had a dormitory upstairs, the Jucos slept in a large room adjoining the dining room downstairs and the officers had a small room next-door. There were beds provided here, which made it easier and more comfortable We spent a lot of time on the beach as it was at the end of the road but had a day trip by coach to Blackgang Chine and Alum Bay where the girls collected the coloured sands. We had a 'Spot the Officer' afternoon when the helpers tried to disguise themselves and walk around the town. The girls, if they thought the strange looking person across the road was an officer, had to ask them the question, "Which end of the bath do you sit?" A list had to be signed off and it was off to find another one. All too soon the week came to an end and we did the journey in reverse but many girls were helped in their Christian journey and friendships made that are still there now.

10.   Billy Graham

IN 1953 MUM AND DAD DECIDED TO MOVE AGAIN. London this time into a three bed roomed flat in Grove Park South East London; Mum and Dad were convinced that they would be able to settle in London especially as a lot of relatives were in South London as well. Dad had got a job in Earlsfield working as a dry cleaner; Pam had been able to transfer from the Royal Aircraft Establishment where she was working, to another civil service department in London Joyce got a job doing secretarial work and I managed to get a job in the photographic department of Boots the Chemists at their head office near Waterloo Station.

Mum and Dad had looked at the flat before they moved in; it was owned by the local council and they thought it was lovely. Unfortunately, it was right by the railway station with the bedrooms that side of the flat and the milk train went by at 3.a.m.! The flats had been built after the war and were very flimsy; Pam woke up one night to hear the sound of quarrelling, and rushed into Mum and Dad's room thinking it was them but they were both fast asleep.

I quite enjoyed travelling on the train up to Waterloo each day and being able to explore London in my lunch hour. The work wasn't that interesting but it was different from Timothy Whites warehouse in Aldershot.

The first Sunday I was there I went to find the Gospel Hall at Downham, Brook Lane Hall. I had been given their address so went along – on my bike, it took about twenty minutes. Everyone was very friendly and there were a lot of teenagers who quickly befriended me and made me feel at home.

I learnt that an American evangelist, Billy Graham was coming to England in three weeks' time to hold a 'Crusade' at Haringey Arena. Singers were needed to form a choir and were being recruited on the next Saturday afternoon. Most of the teenagers from the Gospel Hall were going along so I was able to go with them, Haringey seemed a very long way on the Underground and we had to change trains a couple of times. Although I can't sing very well, I had been in a Christian choir in Aldershot and could just about read music. It was marvellous to meet people from all over London at the arena and to be a member of the choir. Cliff Barrows who was going to conduct the choir sorted us into the different parts of the choir; I was in the alto section. I worked out that I could go straight from work and worked out a route on the underground; I could meet up with the others there.

When the first notes of "Blessed Assurance, Jesus is Mine" rang out across the packed arena, I was hooked. Seldom had anything like this occurred for many years and it had a profound impact on us all. I went along on three nights each week of the twelve week crusade and marvelled as the hordes of people streamed forward at the end to make a commitment to Christ each night while the choir sang, "Just as I am, without one plea--- O Lamb of God I come." There was a coach load from Aldershot there one night but I didn't see them but with sixteen thousand people there it wasn't surprising.

It was that night that Dave finally gave his life to the Lord Jesus. When we left each night there was still a long journey, underground to Waterloo then main line train to Grove Park. At least I didn't have far to walk once I got there. The underground trains were packed and noisy with a great sound of singing of the hymns that had been sung in the arena. We were all quite sad when it came to an end.

I've always thought that we may have lived in London for no other reason that I'd been able to participate in the Crusade, which made it worth living there. Then we heard that a film was being made, which had the Crusade as part of its content, and we had to go along one Saturday afternoon so that the choir could be recorded. It was so much different without the congregation there but we had the added excitement of being outside when Billy Graham, Cliff Barrows and George Beverly Shea came out and our cameras were busy snapping away. I managed to get some good close ups and had to reorder them many times as everyone wanted a copy. The film was called, "Souls in Conflict" and shown at cinemas; I went to see it but was unable to pick myself out on the screen.

11.   Fun at Work

MUM AND DAD ONCE AGAIN DECIDED that they had made a mistake in moving and wanted to move back to the Aldershot area again! The 'Tailor and Cutter' magazine was once again on their newspaper list and Dad started applying for jobs again. I don't know why but they couldn't get council accommodation this time so were going to have to get a job that offered a house too. After a while, I decided to move back to Aldershot so I could be on hand if a house came up as Mum had contacted various agencies in the Aldershot area. I stayed with Mavis and her mother, which was quite convenient but there was nothing around. I was able to have my old job back at Timothy Whites warehouse as a stock control clerk and it felt as if I had never been away. Dave had grown up however although he was still only fifteen and just starting his last year at school – I was twenty! .

Dad got a job working with a big dry cleaner in Fleet who had an empty flat in Farnborough; right in the middle of town above a dry cleaning shop that Dad's new boss owned. There was somebody living in it who was working out a month's notice so we had to live in a caravan for six weeks. This was situated on a farm at Mytchett. There were only two small bedrooms in the caravan and it was a squash – especially as Mum brought her treadle sewing machine with her! Joyce had a job in a shop in Ash Vale, to which she could cycle. Mum still took in sewing and spent a lot of her day at it. Unfortunately, Joyce and I were bitten to bits by the plentiful mosquitoes that shared the farm with us; there had been chicken sheds where the caravan was parked. We were pleased when the flat was empty and we could move in.

The flat was old and dark and very inconvenient with a long corridor, off which opened the two bedrooms. The bath was in the kitchen, covered over with a board, and the toilet opened off the kitchen. There was no washbasin, so we had to use the kitchen sink for a wash. The only other room was huge, with a fireplace at each end, which never warmed the room adequately. There were also two attic bedrooms up a steep staircase but only one had a fireplace. Altogether it was not a very nice place but Mum and Dad were desperate, so we moved!

Pam had decided to stay in London and rented a flat with a girlfriend. She had a good job as a secretary at the Crown Agents to the Colonies, with whom she eventually went to what was then British Somaliland. While she lived in London, she'd come home from time to time for the weekend and we would sleep in the attic bedroom and talk for half the night. Joyce had her job locally and I went back to my old job at Timothy Whites and Taylors. It was a longer bike ride, half of it being through the Army Camp at Northcamp. Mum got a house cleaning job as a change from sewing but used to do repairs for the dry cleaners for whom Dad worked.

The flat was inconvenient and, owing to the size and loftiness of the rooms, was cold and draughty and it was obvious that we wouldn't be there for longer than necessary. I changed my job and obtained one with an electronics firm in Frimley, working in an office with five or six other people. It was totally different to my last job and I was now cycling in a different direction and could go by bus if the weather was bad.

I enjoyed my time there, as it was the first experience I had had of discovering that work could be fun! At break time there would often be paper clip fights and a lot of friendly banter took place. We started a 'nutty' club, with which to buy sweets and often all eight of us had lunch together at a nearby café. One of the things made by the company was an early telephone answering machine; it took up a whole room! We had to do all the costings for the components.

Another girl, Sue, came to work in the office and we quickly became friends. We've kept in touch right through the years, she started coming to the church I went to in Aldershot and, as she lives in Basingstoke now, we see each other from time to time. We would usually catch the bus to Aldershot bus station then walk the last mile to the Church at Park Hall. A girl who lived near Sue started coming to the weeknight games evening but one night Sue was unable to come. Jenny had cycled to get to our flat and we were going to catch a bus. We reached the end of the road as the bus passed; it had to turn a corner before it reached the bus stop, so we started running!

As we ran across the Zebra crossing a car was coming; it caught Jenny and she fell down! She was in a lot of pain; there was a phone box nearby and someone phoned for an ambulance. I went with her to Farnham Hospital but couldn't stay long with her. I stayed long enough to be told that she had a broken leg. I then caught a bus back to Aldershot, told the 'Covies' what had happened then had to walk the mile up to the bus station and catch the bus back to Farnborough. Jenny's house was two miles away so I had to cycle there to tell her mother about her mishap. By the time I finally reached home I was exhausted. Jenny went on to make a full recovery and was soon back at Covenanters with us.

12.   Dave Goes to College

DAVE HAD STARTED ON HIS PRE-COLLEGE farming experience year working on a small farm on Exmoor and loving every minute of it. Despite the age gap we became friends and the postmen were kept busy!

We had a holiday together at Ilfracombe. I managed to catch the right trains as I had to change at Reading, and Dave met me at Ilfracombe station. We had arranged to spend a day at the farm where Dave worked; it took a long time to reach Eastern Ball. It necessitated a train journey to South Molton, a bus from there to North Molton, and then a four mile walk! I had never been anywhere quite so primitive before and also had difficulty understanding what Mr and Mrs Ridd were saying. I had wondered why Dave spoke with a bit of an accent every time he came home and I could now understand why. By the time we arrived back at the Guest House I was absolutely shattered. We went on a coach trip to the Doone Valley as we were both fans of R.D. Blackmore's 'Lorna Doone.' Dave's boss had the name 'John Ridd' and could trace his ancestry back a long way. All too soon the holiday came to an end and we waved goodbye as the train took me back to Aldershot once more.

We moved yet again – I did warn you – to a village the other side of Aldershot to a village again, Weybourne, pronounced 'Webbun' by the local inhabitants. I could walk to the Gospel Hall but it was much further from my place of work, about eight miles. I could get there by bus but would have to change in Aldershot. After a couple of years, I decided to change my job as it was just too far to travel. I obtained a job in the office of a mobile grocery shop and bakery, not very well paid but I had no travelling expenses and it was just a ten minute bike ride. The shop and bakery were as a side-line at a farm that grew hops and they employed a lot of Italians who were descended from those who had been billeted nearby during the war. They all seemed to be related and were known collectively as 'Luigi'. Some of them spoke no English, they had a collective 'book', which was filled in every time they came in for groceries but I don't remember what the system was of charging them. I worked mainly with the bakery section making up books with the aid of a 'Ready Reckoner,' writing out invoices, and manning the telephone. There was one difficulty with being out in the country and that was that there were mice around! One even found its way into a loaf of bread and was baked!

Mum and Dad settled in Weybourne for some years. Mum had been cleaning for a lady in Farnborough so carried on travelling there every day, catching two buses to do so. Dad decided to change his job and found a job as a storeman at a Ministry of Defence Establishment but had an accident and injured his knee. He had to give up his job as a storeman so went back to dry cleaning again. He received compensation for it, which enabled them to buy their own house.

When David finished on Exmoor he was thrust into a very different environment at Sparsholt Farm Institute, as it was then, it is now Sparsholt Agricultural College. When David went there it specialized in skills needed in a farm environment but now it is into the leisure industries with courses on Green keeping, fish farming, horticulture and small animal courses. Our granddaughter, Andrea starts there next September doing an animal management course. Dave found life very different, especially the food, which was adequate but not like Mrs Ridd's cooking. It was strange too to be amongst so many people after the solitude of Exmoor. It did have the advantage of being only twenty miles from Aldershot so he was able to come home most weekends, usually hitchhiking.

There was a farm shepherd at Sparsholt to teach practical Shepherding skills, but he died when the ewes were a month away from lambing. The students had a rota system for covering the extra work and during lambing time had to sleep in the shepherd's hut out in the lambing field. This looks like an old shed on iron wheels with a chimney coming out of the roof! There was usually an old iron stove inside with a saucepan bubbling away for the endless cups of tea.

Dave liked the early morning shift as he was good at getting up and at three o'clock he was wide awake if he was needed. He was called out by the other students or the shepherding tutor if there were any difficult birthing problems as he had more experience than any of the others. He was top student while he was there and was awarded the National Certificate in Agriculture with Distinction. He took his driving test in a borrowed Landrover while he was there and passed, after spending the night with the sheep.

Years later he returned to Sparsholt on a day release course in Farm Management and was awarded a Full Technological Certificate.

13.   Our Wedding Year, 1961

As THE TIME CAME FOR HIM TO LEAVE Sparsholt he started answering advertisements for shepherds; he had several replies, one of which was down in Kent and the advertiser wanted somebody with milking experience as well as shepherding skills, to do relief milking. It was a long way from Aldershot but there was a cottage available too although it would be another two years before we were able to take advantage of it. He travelled by train, being met at Ashford station by the farm foreman who showed him around the farm. The milking was done using a mobile milking 'bail' this had four stalls where the cows would stand to be milked. It was moved daily to the field that the cows were in, they were getting rid of the herd over a period of time so didn't need a full time cowman. David then had an interview with Mr Older, told that he would be contacted and subsequently was given the job to start at the beginning of July.

Harvesting was just beginning and with a large acreage of corn David had to help with the tractor driving. There were several other men employed, some did nothing but tractor driving but Dave would be relieving the cowman when he had his days off as well. Dave managed to find lodgings in a village about two miles away so was able to cycle to work each day. There were three other chaps living there working on other farms and they all got on very well together. The landlady loved to look after these young men; her husband worked in a grocer's shop in Ashford, eight miles away; he too cycled to work. They attended the little Baptist Chapel in the village and Dave often went with them. We were engaged when Dave was eighteen and a half and married two years later. Our house was a Kentish tile hung, brick building with three bedrooms, sitting room and kitchen complete with a black leaded kitchen range, which I had to learn to use – and black lead! It was replaced eventually by a Rayburn, which was much more efficient and cleaner.

We bought mainly second hand furniture to furnish our house. We found a four piece bedroom suite advertised in the paper for twenty eight pounds. It was made of walnut and had been bought new by the sellers in 1924! (We are still using the dressing table and small wardrobe). We did buy a new table, four chairs and a cupboard for the kitchen also a baby Burco boiler. I bought a second hand drop head wringer and with somebody's cast off armchair, had furnished the kitchen. The sitting room had a three piece suite that David had bought second hand, my treadle drop head sewing machine and a small second hand radiogram. Mum and I made all the curtains, some were made from ones that Mum or Dave's Mum had finished with. I bought new ones for our bedroom to match the bedspread and eiderdown and felt very extravagant.

We decided to get married in May 1961 when Dave would be twenty. Mum made my wedding dress and the dresses for Dave's two sisters in an apricot taffeta and my cousin Doris in sea green, as they were to be bridesmaids. My headdress and veil belonged to my sister Pam who was married in British Somaliland where she was now living. She had gone there with the Crown Agents to the Colonies, met and married her husband who was in the British army there. We arranged the catering ourselves for approximately one hundred and I had my Covenanter girls to act as waitresses; they also volunteered to wash up. The person I worked with made the three tier cake although one tier I cut up and boxed ready for Mum to post to absent relatives and friends who couldn't come to the wedding. We were going straight down to Kent after our honeymoon in the Lake District and we thought it would be easier for everyone if we did as much as possible before the wedding. I met Dave's Dad in Aldershot two days before the wedding.

"I've just been to the dentist Eileen."

"Did you have to have anything done?"

"Yes, look. I had to have some front teeth out!"

"Pop! Everyone will notice them!"

"How can they? They're not there any more!!"

We were married in The Gospel Hall where we attended each Sunday; with David's uncle who was a missionary in Malaya performing the ceremony. There were a hundred and twenty guests at the wedding and we used the schoolroom at the back of the Gospel Hall for the wedding reception. It was a simple meal of sandwiches and cakes but everyone seemed to enjoy it, the photographs taken at the reception showed the sandwiches curling up as we hadn't covered them! We borrowed Dave's father's car, a little Ford Popular to go away in, and travelled via London as I wanted to give 'Auntie Rice' my bouquet and then we trundled up the M1 to Birmingham. The car had been fitted with a new engine not long before and we were restricted to forty miles per hour. We dozed in the car somewhere on the way, and arrived at our destination in Bowness-on Windermere in time for breakfast. We even went to the morning service but slept most of the time.

We had a washbasin in our bedroom but had to share the bathroom and toilet with the other residents of the bedrooms on our floor. I went to use the toilet one morning and met a man in his pyjamas and dressing gown coming out carrying a chamber pot.

"Good morning, lovely day isn't it?"

"Yes isn't it." I mumbled feeling very much at a disadvantage.

We stopped overnight with friends on the way home as Dave was taking part in a shearing competition the next day on the Romney Marsh. As he was still only twenty he had entered the 'under twenty one class' – and won it!

I found it quite difficult at first living so far away from the family and as Brabourne, the village in which we now lived was only a hamlet, it was hard to get to know people. In fact, I don't think I was really accepted until lambing time when I had an orphan lamb to rear. Larry followed me everywhere and came with me to the little shop and did a wee on the floor. Baby lambs have a real 'aah' feeling about them.

Dave answered the door one evening to find a small boy standing there with a squirming cat in his arms.

"Mum says will you do the cat?" He thrust the animal into Dave's arms and promptly ran off!

Dave shut the cat in the bathroom and went next door to ask advice of Alf, one of the tractor drivers who lived there with his wife, two sons and eight year old daughter, Shirley.

"Alf, what am I supposed to do with a cat?"

Alf laughed, "'Tis easy, last shepherd allus castrated the village cats."

Dave went back indoors and thought the matter through; he was used to castrating lambs but how was he going to hold the cat?

In the end, he put the cat down a coat sleeve with the necessary part of its anatomy protruding and with his sharp shepherds knife performed the operation. He put the coat down and the cat made its escape through the open door. Dave thought it had gone for good but a couple of weeks later he saw the little boy again, "Hey mister, Mum says to tell you when I seed you that cat come home O.K."

14. A Sad Loss

WE HAD QUITE A LOT OF VISITORS and I was glad that we had the spare bedrooms. Mum had given me the two single beds that used to be in my bedroom at home and they fitted nicely in the smallest room. We bought a second hand double bed for the other spare room and utilized tea chests and wooden boxes for bedside tables.

Once a week I'd write out a shopping list and Dave would take it to the grocer's shop in the next village with the payment for the previous week's shopping. Two days later the groceries arrived in a box, which would be carried in and placed on the table. If I ran out of anything there was a small shop in our part of the village, which stocked basic necessities.

Dave's parents and his sister Maureen came to stay. Dad was very useful and loved doing odd jobs. He went into the front garden to cut the grass down, as it was too long for a mower to cope with. He had been out there a few minutes when he came in carrying his grass hook.

He said, "I've cut my hand" and promptly fainted!

Mum had been a nurse and when he came round she bound up his hand.

"You're going to need stitches in that I think Nat. Can we get hold of David to take him to Outpatients at the hospital?"

I ran down the road to the farm and luckily David was working in the dryer.

"Dave, Dad has cut his hand and needs to have it stitched. Can you borrow a vehicle and take him in to Ashford Hospital?"

"I'll see if I can borrow the Land Rover, then he can lie down."

He went off to get it while the rest of us had the inevitable cup of tea. We all needed it for the shock! Dave put a blanket down in the back of the Land Rover; Maureen offered to sit in the back with him. They were just about to set off when Mum, who had decided to stay home with me, thrust the washing up bowl into Maureen's hands.

"In case Dad's sick!"

They arrived home, Dad with a neat bandage around his hand and his arm in a sling, looking very white.

"Are you all right Nat. Were you sick?"

"Yes he was," replied Maureen. "He lost his false teeth and I had to fish them out!"

I found out that my neighbour Mrs Shorter and her friend had never been to London, so in the school holidays I arranged to take them on the train. Shirley, Mrs Shorter's eigh- year-old daughter, came too so we arrived in London about ten o'clock and began our sightseeing trip with a 'cuppa' at Lyons. They were all eyes at the waitresses with their black dresses and frilly white aprons. We then had a ride round London in an open top bus and managed to get the front seats. We saw the Monument, St. Pauls Cathedral, the Old Bailey, Piccadilly Circus, and all the places in between. We then had a snack and fed the pigeons in Trafalgar Square before walking along the Mall to Buckingham Palace. Then it was another cup of tea and a cake in St. James Park before catching the homeward train to Ashford, where Dave was waiting to meet us with the farm Land Rover.

After we had been married about a year, we decided to buy a van. David's employer had a scheme whereby he'd lend any of his employees' money to buy a vehicle and would take the repayments directly out of their pay packets. There was no purchase tax charged on vans, so it worked out a lot cheaper than a car. We had managed with buses up to then, which were very infrequent from the village, or else would cycle four miles to the main A20 road, where we would dump our bikes under a hedge and go by bus the rest of the way. Sometimes other bikes would have joined ours; we never even had a padlock and chain on them and they were always there when we returned for them. We worshipped at a Gospel Hall in Hythe, where we soon found good friends. One couple ran a children's nursery in their house and we often went there for lunch on Sundays. They took in young children from the council and usually had about twelve there at any one time. We often took them out along the promenade; it was lovely to be so near the sea.

Mrs. Bloomfield was a lovely motherly person whom the children could really relate to. She'd sometimes take some of them out in her little Mini car for a drive to see things of interest. One day she took them for a drive in the Romney Marsh area, which was crisscrossed with dykes or ditches. She half turned to point out some sheep with their lambs in a field to the four children in the back seat when she found the car careering down the bank of the dyke. Somebody saw her predicament and helped her to get the children out; there were no seatbelts or restrictions in those days. Nobody was hurt but she always made sure that she had another helper with her after that.

I became pregnant and the baby was due at the end of February. Dave was starting lambing on March 1st. Two days before Christmas I was feeling very lethargic and the baby seemed to weigh a ton. One afternoon after Dave had gone back to work after dinner, I began to feel ill and lay down on the settee. There were no mobile phones in those days and the neighbours were out at work As soon as Dave came in he realized that something was wrong. He went down the road to the phone box and rang the doctor who arranged to see me at half past seven at his surgery. We had our tea of bread and cheese although I didn't eat much and set off to the doctor's house where the doctor examined me and said that he'd like a second opinion and that the consultant lived not far away. When he came and examined me he said that I had to go to hospital right away.

We went home to pack a case and I remember I gave Mrs Shorter a chop for Dave's dinner the next day. We arrived at the hospital where I was admitted to the Maternity ward, which I was surprised about as I thought one only went there to have a baby! I had to take enormous pills and felt very strange. Early next morning, I gave birth while fully conscious, to a stillborn baby and couldn't seem to make anyone understand that they would have to tell David! What I didn't know was that he had already been told when I was admitted that the baby was dead and that if I didn't have the baby soon, I'd be dead too!

I had given birth in a ward with only a curtain separating me from a woman in labour, whom I was told had had a still birth a year before and had just had a live baby girl. It didn't help at all.

It was three days before Christmas with deep snow on the ground. I was kept in for another fortnight before I was allowed home and some evenings Dave would be the only visitor in the ward. I was given a bed in the post-natal ward and had the agony of seeing the other women with their babies who were brought to them for feeding.

I desperately wanted to see Mum and Dad so once I was feeling a bit better while I was still in hospital, Dave set off to get them. He travelled to Aldershot and was there by the evening, he stayed there the night and brought Mum and Dad for what was a short visit as my blood pressure kept going up. He took them back, spent another night in Aldershot before arriving back in Kent. When I went for the six week check-up I was told not to conceive again for a year, which I found rather upsetting.

15.   A Whitsun Trip

WE HELPED AT A YOUTH BIBLE CLASS at the church and decided to
take them away for a weekend with another group from Aldershot. We contacted a farmer near Bognor and arranged to go for
Whitsun as the weather would be warmer for sleeping in a barn.
Dave managed to borrow a lorry to transport the ten boys; this
was usually used to take farm workers to work and it would
double up as a bedroom for Margaret (who was coming from
Aldershot with Roy), and me. We borrowed a camping stove
and as many pots and pans as we could find and set off on the
Friday evening. It took us about an hour to get there; Roy and
his party had already arrived. Everyone went for a look round
the farm while Margaret and I tried to set up a kitchen in one of
the barns. It was a bit primitive but it worked quite well although we had to rely on having the water supply topped up
from the farmhouse. We just gave the boys soup and bread that
first evening with thick wedges of cake to fill up the gaps.

It was cold! There were tarpaulin sides and top to the lorry
but nothing at the end. Margaret and I heaped coats and
anything else we could find on top of us. It was a short night and
we were frozen and dog tired in the morning. The boys fared
better as they were in the hay barn, which was quite warm. It
was cornflakes plus baked beans on bread for breakfast and it
was decided to spend the day at Bognor, which was about ten
miles away. Margaret and I were going later when we had
prepared the sandwiches for lunch; we were going in her car so
waved the lorry load of youngsters off. We had a cup of tea

before starting on sandwiches for fifteen boys either ham or cheese plus crisps, cake and an apple. We arranged to meet the boys at the miniature railway station at the end of the promenade, which was the only place in Bognor that Dave and I remembered. It didn't take us long to do the sandwiches and we decided to have a wash before we went out. We heated a kettle of water and found a shelf on which to stand the bowl. Margaret had just taken her jersey off when we heard somebody coming. We both dived for cover; Margaret hit her head on a bit of machinery and the blood started pouring out! As long as she kept a towel pressed on it, the blood stopped but we realized that she would need to have it stitched.

Somehow I drove her car to the farmhouse about quarter of a mile away – I hadn't yet started driving lessons – and ran in to ask the farmer's wife what to do. We decided that Margaret ought to go to the casualty department at Chichester Hospital and it was another half hour before we were able to use one of the tractor drivers to drive us to the hospital. We put the lunch in the car, hoping we wouldn't be too long at the hospital. Margaret was as white as a ghost and on the verge of fainting. The time for meeting up with the boys was past but Margaret was soon in and out of casualty and our driver was willing to take us on to Bognor.

We drove to the miniature railway station but *it wasn't there!* In its place was Butlins Holiday Camp! I walked the length of the promenade, having arranged with the car driver to meet up with him at the pier – at least *that* should still be there. At last I found them. It was now two o'clock and everyone was starving. Margaret and Roy went back to the farm so that Margaret could rest. The remainder of the weekend passed off uneventfully but before we left, Mr Humphries, the farmer, asked David if he would manage a small farm that he had bought. He also offered Roy a job on the main farm as he had always wanted to work on a farm.

16.   Farm Manager

DAVID HAD BEEN AT HIS CURRENT JOB for four years now and it was too good an offer to refuse; Roy too felt that it was a chance to do something that he really wanted to do, he had been working for the GPO in the telephone installation department since his demob from the army.

The farm Dave was going to manage was ten miles away from the main farm and the house although lovely, was very primitive and isolated. It lay back from the main Petersfield – Chichester road with a big vegetable garden in front. Roy and Margaret had been allocated a house near the main farm at East Dean; ours was near the village of Chilgrove. We needed a bathroom installed as the previous tenants had used a 'bungalow' bath in an outhouse; the kitchen had no sink although there was a shallow one in the scullery, and only an antiquated range to warm the kitchen. There was no hot water, we had to go to Margaret and Roy's when we wanted a bath and the loo was in an outhouse up the garden and had to be emptied when necessary.

It was about six months before we finally felt settled there, we were glad to have an indoor toilet, it was cold and dark when we had to make the trip up the garden in the evening! We were now able to have friends and relations to stay; Dave's parents and his sister came for Christmas and it was nice to have so much space. We had four bedrooms, a study, sitting room, scullery, kitchen and dining room. We had a Rayburn now in the kitchen, which heated the water. The large bath room with separate toilet, had been built in a bricked up room behind the dining room fireplace. The house was a couple of hundred years old with lots of beams and low ceilings.

I decided to have a duck for our Christmas dinner and Dave bought it from a local farm so it was nice and fresh. I had looked in my cookery books and discovered that an orange sauce was the thing to have with it and sage and onion stuffing. Dave's parents were coming in time for lunch, which was when we were going to have our Christmas dinner. I got up earlier than usual so that everything would be ready in time. Dave had to feed the sheep but he peeled the potatoes for me, which was a help. I made the stuffing and left it to cool down before filling the bird, the sprouts were very fiddly to do, – no frozen ones in those days – I had made a Christmas pudding and put that on to boil. I couldn't put off stuffing the duck any longer or else it would never be ready. I got a handful of stuffing and pushed it into the cavity and continued stuffing it until no more would go in. I put it in the roasting tin with some lard and it went into the oven. Dave's parents and Pat arrived so I made coffee for us all.

I laid the table, it looked quite nice with a red and white checked tablecloth and my 'best' china with yellow roses. Dave carried the duck in on its matching rose strewn platter and I handed him the first plate. The duck looked mouth-watering and as Dave started carving the first slices from the breast for his mother, it smelled gorgeous. "Is that enough for you Mum?" "That's fine Dave." Would you like some stuffing too?" He delved into the depths of the duck with a spoon and started to look puzzled and reached into the cavity; he pulled his hand out of the bird and held the bag of giblets up! We all had a good laugh but I never made that mistake again.

The dogs loved it at Hill Lands; there was plenty of space in which to run around. Jill developed the knack of 'moling' She would stand near a mole hill, head cocked on one side listening. She would then start to creep along the ground, typical sheep-dog fashion and stand, one paw in the air and then pounce and dig furiously; occasionally she was rewarded and caught 'Moley' in his black velvet coat but not often.

The farm was mainly arable and pigs with quite small fields but a hilly terrain; the views were marvellous situated as we were on top of a hill. In our scullery there was a small hole in the concrete floor, which was the only opening of a well, which must have been quite deep owing to our situation. We often had mice in the large pantry and had to make sure that everything was covered up. One day I found one dead in a pot of home-made jam! There was plenty of fruit around with apples and pear trees in the garden and blackberries grew along the sunken lanes so I was soon making jam.

17.   A New Arrival

I BECAME PREGNANT AGAIN; the baby was due at the end of September. All went well until June when my blood pressure started to rise. The doctor sent me home with strict instructions to rest as much as possible. This was hard as the house was so big and the garden needed a lot of attention although with the hot weather I felt quite lethargic.

A week later the doctor was even more insistent with the threat of hospitalization if my blood pressure was still climbing. I worked out a routine; Dave brought me a cup of tea in bed and went off to do the necessary feeding returning for breakfast at nine. I'd be up by then, cook his breakfast, wash up and go to bed again for a couple of hours when I'd get up do some washing and get a meal. My after dinner nap turned into an all afternoon one and I'd not get up until five thirty when I'd get us another meal. I'd relax in the evening with my feet up and then be in bed by ten o'clock! It was in vain as the next week the doctor sent me to St. Richards hospital in Chichester on complete bed rest!

Dave was harvesting so couldn't get into visit me until late in the evening. Visiting was very strict there, only two visitors per visit and no changing over, they had a list and ticked visitors off on it, but they made a special allowance for Dave who visited after he had finished work, usually about nine o'clock, and even supplied him with cups of tea. I was in a ten bedded antenatal ward, which meant that on the whole, people were coming and going the whole time. Those who were waiting for babies, were often given premature babies to feed as some mums had to leave the babies behind to grow a bit but as they themselves were fit, they had to go home. At one time however, I was the only person left in the ward, occupying the furthest bed from the door. Mums would come in and a couple of days later have their baby and go on to the postnatal ward. I became quite depressed, as I couldn't read much because of the medication that I was taking, and it was before TVs were available for hospital wards. When Dave came in one evening he went to see the Sister and told her that either she'd put somebody else in the ward with me or else he'd move in himself! I soon had company again.

I then had the company of a couple of mums whose babies were in incubators little knowing that in a couple of weeks' time; my baby would be in one. I went into labour about five o'clock one morning; the sister asked whether to call David, but as she thought I'd be some time, we decided to leave it as he had only left me at eleven o'clock the previous evening. My waters broke and she wheeled me into the delivery room but discovered that things were happening faster than she had thought. She had to phone for the doctor to come and let Dave know too. He was there in time to hold my hand for a while but had to leave when the team realized that I would have to have an episiotomy and forceps delivery as my contractions were not strong enough. He waited outside the door, heard Andrew's first cry and saw him in his incubator when it was wheeled past him; then he had to go to work!

I didn't see Andrew until later in the day when a nurse took me in a wheelchair to visit him. He weighed three pounds twelve ounces and was in an incubator; he looked so tiny with various tubes everywhere. I was moved into the post natal ward where it was much busier but I was still on bed rest because my blood pressure wouldn't stay down. I had to express my milk using a breast pump as Andrew was being tube fed because he was so small, but at least I could go and sit by his incubator for a while. Gradually we both got better and I was able to go home but Andrew had to reach the magical weight of five pounds before he could come home. This took another month; I still expressed my milk with Dave doing the 'milk run' to the hospital each day. The last day and night that Andrew was in hospital, I had to stay at the hospital to make sure that I could feed him properly myself. We passed the test together and then we were on our own.

He was continually sick and when the district nurse saw him she felt that he was getting too much milk as I was breast feeding and advised me to hire some scales from the chemist and weigh him before and after a feed. This was a palaver but apart from me being constantly lop sided as the only way I could give him enough 'sucking time' was to give him one breast at each feed. We had a little room leading off from our bedroom and we kept an electric heater on all the time for Andrew. It was quite cosy when he needed a feed in the night, which he did for a long time after we brought him home.

Years later when Peter was a baby, I took him to the baby clinic and the nurse expressed surprise at the way that Peter was sick. (He was sick into the scales pan as he was being weighed.)

"I've never seen a baby being that sick before."

"I have; his brother!"

There was never any explanation as to why the boys were both so sick, Andrew put on weight slowly and did most things that he should be doing like standing and walking much later than other babies of the same age. He was quite a miserable baby but I'm sure that he was in pain some of the time because he had developed a hernia. Even after he had had an operation to sort it out, he still cried a lot. My friend Elizabeth who lived in Chichester and attended the same church as us, was a great help. Her two boys were a bit older than Andrew and she would often look after him while I caught up on my sleep. She employed a nanny so one more child didn't make much difference; it was certainly nice to have a 'bolt hole'.

Other people in the church had been very kind to David while I was in hospital. If anybody wanted to know what they could do for David he usually invited himself for a meal, usually leaving with a cake as well. One elderly lady was rather taken aback when she offered to give David a meal.

"I'm bringing pigs in to market tomorrow morning; I'll be in for breakfast at 8.30.a.m."

She cooked a full English breakfast for him!

We went to church in Chichester and found a welcome at the Gospel Hall. It was a big church with a lot of teenagers, and was quite near to the cathedral. After the usual service on Sunday evenings an altogether different type of service was held for the teenagers. It was in the 'teddy boy' era and weirdly dressed youngsters would come in. Some people would go out 'fishing' and invite people in off the streets. A speaker would be invited who would be relevant to the youngsters. The first week we were there they were expecting Fred Lemon, a man who had been converted to Christianity while he was in prison.

Dave had had his hair cut the week before we moved and had opted for a crew cut – he was mistaken for Fred Lemon, the speaker, but he soon put them right.

We soon got to know people and could invite them back to the farm; we became especially friendly with two girls, Jill and

Marilyn who were in training to become nurses. As I had four bedrooms, I was able to keep two beds made up for them and they often came out to stay, which was especially useful when I had Andrew, as they would baby sit for me so I could get some sleep. In fact when Andrew had his hernia operation Jill had him in her room for the first night he was home. On another occasion when Marilyn was staying I drove into Chichester leaving her to get Andrew up and dressed. When I came home she was doing some baking with Andrew laid on the table fast asleep along with the flour and sugar.

I usually took Andrew with me when I did my shopping in Chichester. I somehow lifted the pram into the back of the van – with the hood down, the pram brake on– and drove the ten miles. Getting the pram out was easier and there was often a man in the car park that would help me. I had a nasty experience one day in the car park. Barriers were just coming in and lifted as they were approached. I was too close behind the car in front and the barrier started to come down when I was under it!

Chichester was a nice place to shop and Sainsbury's had just opened up; the first supermarket to do so. They didn't allow prams inside; we had to leave them outside on the pavement. It was the same for most of the shops and we didn't worry about it. It was quite common to find someone rocking the pram as Andrew had a loud cry! When he grew older I bought a car seat for him with a little steering wheel, he loved sitting in the front of the van with me. It would fail every current safety test; it just hooked over the passenger seat with nothing else to secure it to the van but he didn't come to any harm.

A family of four lived about a quarter of a mile away up in the woods. The two girls caught the school bus at our garden gate so sometimes their mum would bring them in for a cup of tea and a bit of 'baby worshipping' on their way home. There were no milk deliveries where we were and Sainsbury's didn't stock milk, but we were able to get some from a farm half a mile down the road that had a dairy farm. I bought a couple of two pint 'churns' that would fit in the fridge, and it made a nice walk with the pram.

18.   Weekend Camp

WE BECAME INVOLVED AT THE CHURCH with the work that was being done with the young people and arranged a weekend camp for them. We had friends at Storrington who had suitable barns as we were taking a mixed party this time so would need at least two barns. We managed to borrow a caravan for Andrew and me and eventually the date came round and everyone made their way to the farm. It was before the age of blow up beds so everyone was asked to bring a mattress-sized cover in which to stuff straw. What fun they had doing it and more than one straw fight ensued. Andrew wasn't walking but loved standing up and we had found a round wooden cheese box, which I covered inside with sticky backed plastic and he was very happy in that. I could put toys in it, it was very stable and he could reach the top of the sides and haul himself to his feet. Everyone loved him and there was always someone to keep him amused.

We had a trip to Bognor on the Saturday and loaded up the cars to overflowing. We went to the park there where some played miniature golf while the younger ones liked the playground. When they arrived back I had to start on the evening meal of sausage and mash, we had managed to borrow a calor gas cooker so had about four rings and the oven to use. At this camp there was only about twenty youngsters but a few years later we had as many as eighty. The youngsters had peeled the potatoes in the morning, a big Dixie full, which took some time to cook; we had baked beans to go with it. We had managed to buy some big tins of creamed rice for pudding so emptied them into a saucepan and that cooked while we were eating our first course.

By the time it was all ready and the potatoes had been mashed everyone was really hungry. We had bought a couple of big bottles of tomato ketchup and soon they were all tucking into their meal. Over the years we had everything down to a fine art and even managed a roast dinner for them most years, we never attempted chips though. It became traditional to serve rhubarb and custard for pudding at the last meal. All the farmers on whose farms we stayed seemed to have enormous rhubarb beds and they offered us as much as we wanted. We had to empty the loos, so thought it was sensible to serve it at the last meal!

After washing up we all met again in the barn and spent an hour singing silly songs and performing sketches, it was surprising how much talent there was at camp. They all loved songs like 'Green grow the rushes-O', 'Three wood pigeons' and 'Jingle puffer' and as the evening grew darker it was a very evocative scene. We kept this format for the first evening right through our subsequent years. We finished up with a short 'God slot' when we sang some Christian choruses and Dave talked to them about God's love for each one of them. We still meet people today who remember the times they spent at barn camp.

19.   Sheep Shearing

DAVID STAYED AT HILL LANDS for a couple of years but felt that he'd like to get back with sheep; he also wanted to do shearing for a season He felt that he was now in a position to try for a manager's job at a bigger farm. He handed in his notice and the boss decided to offer Roy, David's friend, the vacant position. Dave started reading all the job vacancies in the Farmers Weekly and put a notice in to see if he could get shearing work in the South. We also started house hunting as we would need somewhere to live while he did the sheep shearing.

The shearing jobs started coming in; Dave went to see the bank manager for a loan with which to buy a shearing machine and engine. The bank manager seemed happy to oblige with a loan of £100.00. Somebody gave Dave a small trailer on which he mounted the machine; he also had to buy a grinder with which to sharpen combs and cutters.

He then saw an advert in the *Farmers Weekly* for a farm manager, so decided to apply for it; he received a reply asking him to attend for an interview at an address in Hurstbourne Tarrant. We had to look it up on the map; it was just north of Andover, Hampshire.

Elizabeth had Andrew for the day, the interview was quite early, 10 o'clock, but we managed to reach there in time. David had the interview by himself and then the farmer and his wife wanted to meet me too. We were taken to see the house in which we would be living if he was offered the job, which was actually two houses that had been made into one. It was quite big with four bedrooms and plenty of room downstairs. It looked like a nice village as we drove through it but we were not sure if Dave would be offered the job as there were some other applicants. After a week we were beginning to think that Dave hadn't got the job. I picked up the post and took it into David who was eating his breakfast. "Quick, open it!."

David never rips open letters but slits them with a knife – so after wiping the marmalade off it first – he read through the letter,

"The job's mine but they would like me to start on the first of July because of the harvest and they would like to discuss the problem of me wanting Sundays off. I've got one flock of sheep that need shearing on the third but if they are willing to change, it would be all right."

We were going to have to move when Dave finished at Hill Lands while he was doing the shearing, but we found a two bed roomed flat to rent at Selsey, we knew the residents in the downstairs flat and they had a young baby too. We had a lot of sorting out to do as some stuff would have to be stored in a friends' barn, the flat was partly furnished too. A big room in the house was also used as a church, at which Dave had preached, and we knew the owners quite well.

We soon settled down and Dave started shearing. It was a good thing that we were in the flat as Dave often had to stay the night if he was at a distance and it meant that I didn't feel alone. One job was at Horsham and David was going to stay with Elizabeth's mother, a real 'matriarch'. He had been shearing all day and arrived there about seven o'clock one evening. She soon had a cup of tea for him.

"Would you like a bath before you have a meal?"

"That would be lovely."

She shepherded him upstairs to the bathroom. "Now, if there's anything you want, just shout out."

"Could you possibly wash my back for me please as it's so itchy?"

"I'd be only too pleased to!"

20. Hurstbourne Tarrant

THE THREE MONTHS PASSED QUICKLY and we were soon packing again. David's new employer had sent a lorry for our things and it took just about two hours to drive to Hurstbourne Tarrant. Dave's parents had come from Byfleet to help us and my job was to decide where all the furniture was going and which curtains belonged to which window. It was a lovely sunny day and Pop took Andrew for a walk in his pram. He came back an hour or so later.

"We've been to a Flower Show; there were classes for everything; there were stump rooted or long carrots, bottles of wine, many different classes of flowers, and the cookery items made me drool. Shall I make a cup of tea?"

In years to come I won prizes myself at the Flower Show and even won a silver cup, my entry one year in the sewing class was a shirt I had made for David and I also won something most years in the cookery class.

We soon settled in but Dave found that managing a big farm was quite a challenge. The sheep were in a very poor state and a lot had to be culled. He instigated the new regime of working a bit later in the evenings so that Sundays were free and the men worked on a rota system to enable the three of them to have every third Saturday completely off in turn. Starting in the middle of harvest was hard but David soon had his system up and running and the men found that they were earning more overtime money and enjoyed their Sundays off. The dryer would often be running long into the night and I can smell the hot, dry smell even now.

We went along on Sundays to the Gospel Hall situated in Vernham Dean, a small village four miles away where there was a congregation of about twenty. We met another young couple there, Nigel and Marion who had a child, Debbie, a little bit older than Andrew. Nigel owned a farm about ten miles away and we were soon great friends and still are. Marion went on to have five more children only one of which was a boy. Marion came from New Zealand and they eventually went out there to live.

David was able to renew the friendship some years later when he was awarded a Nuffield Scholarship to study sheep farming in Australia and new Zealand

Nigel had only been a believing Christian for a short time, which happened when he met a missionary on the boat coming over from New Zealand who led him to trust the Lord. We received a great welcome from everyone that first morning and soon settled in. David took over the boys Bible class on Sundays and went on to start a weeknight activity evening. At one time there were six boys in my small living room taking a lawn mower to bits! Later on they amalgamated with the girls' class and started going swimming in Basingstoke or Swindon as there was no swimming pool in Andover. Dave would hire the village hall for games nights and we were always trying to think of interesting things to do.

21. Another Arrival

WE HAD BEEN IN HURSTBOURNE TARRANT for eighteen months when Peter put in an appearance. I went into labour early one morning so after 'phoning the Doctor, Dave also woke the neighbours who said that they would look after Andrew who was still asleep. David left me at the hospital, thinking that it would be some hours before Peter arrived. He had other ideas though. No sooner had the Doctor arrived than Peter was born – bottom first! This meant that once again I was deprived of that first cuddle of my baby, who was whisked away to be cot-nursed for twenty-four hours because of his traumatic arrival. Once again, I had an episiotomy and forceps. He was totally different from Andrew but extremely hard work owing to the constant sickness. Even when he was toddling around, he was still sick. I had to stay in hospital for twelve days before the Doctor would let me go home. Andrew came with Dave to collect us both but he was more interested in a lorry that he had been given than in his baby brother.

I helped out in the Sunday School where about twenty five children gathered each Sunday afternoon, I taught the twelve pre-school children in the kitchen. The original church was pulled down soon after we moved into the area. It was a 'tin tabernacle' type of building, erected in 1902 and was long past its sell by date. The pews had one long backrest at adult height and children often slipped through the gap. A new modern building was erected with inside loos, a modern kitchen and large room in which to hold services. New interlocking pale blue chairs were purchased and blue velvet curtains hung at the windows. The building has been used a lot and the local pre-school use it now during the week.

Mr Bert Edgington who was the Sunday School superintendent, lived very near the Gospel Hall and was the local baker, living with his wife and sister in law behind the bakery. He always had a bag of toffees in his pocket, which he would distribute to the children after Sunday School. He maintained that "It's like scattering chicken feed – the chickens will come running." He was a lovely Christian and looked after his disabled wife, Ruby and her sister, Olive, as well as running the bakery business and shop; he also had a baker's round and would go to the outlying villages with bread. He was always cheerful and ready to help.

One day I was helping out as Olive was in bed, unwell, and Ruby by now was in a wheelchair, and the lady who helped in the shop wasn't allowed to serve petrol, which was also sold. I also had Peter, a four year old with me. A lady came for some petrol so Peter and I went to serve her. I put the nozzle in the hole and set the pump going. When the tank was full, I pulled out the nozzle *but forgot to let go of the trigger.* The petrol went everywhere, all over the customer and Peter! I sent him, screaming into Wendy in the shop while I coped with the customer who was very nice about it! I'm afraid it left me a profound fear of filling our car with fuel – even now.

Sometimes; we would have a 'night hike' for the teenagers as a change from their usual activities. We usually played a 'wide game' had indoor games of rag hockey etc. sometimes took them swimming but the night hike was the last evening before the summer break. They would leave Vernham Dean about eight o'clock in the evening after leaving tins of baked beans and sleeping bags with me. I gave them a couple of hours then set off in the car to the farm where we were spending the night and heated up water for a hot drink. They had travelled about eight miles across country and eventually arrived and flopped out on the straw. I had a big tin of cakes and after feeding them they settled down for a sleep in the barn, I slept in the car. The next morning I heated up the beans, which, with a slice or two of bread plus a drink, constituted their breakfast. Some boys had Saturday jobs so I took them home in the car. The farmer let the others use his swimming pool before they set out to walk home although, having returned with the car and another car and driver, we eventually ferried the boys home. We still meet boys today who remember vividly the night hikes and barn camping.

22.   Spring Camp

WE DECIDED TO TAKE THE TEENAGERS camping but having no equipment, we thought we could have use of a barn somewhere not too far away because of the transport problems. It was decided to hold it over the Spring Bank Holiday weekend when the weather would be warmer and there'd be the bank holiday. We went that first year to Marion and Nigel's farm; Andrew, Peter and I slept in the house. Another group was joining us so there was about thirty teenagers and six helpers. Dave had bought six chickens for one dinner but there was a problem – they were still alive! On Saturday morning before they went out for the day, Dave got four boys to help him kill and dress the birds. He showed them how to wring their necks, take out the 'innards' and clean it. They all watched carefully as he demonstrated the first one, the chickens had to be taken out of the crate one by one. Dave was helping one boy when another came up to him with a live, half plucked chicken in his hand.

"Dave, I pulled its neck like you did and started plucking it but it jumped off my lap!"

Dave roared with laughter and dealt with it for Adrian.

We had a trip to Swindon by pantechnicon as the other group from Woking had arrived in the furniture lorry. No seats so everyone sat on the floor; it was very claustrophobic but nobody worried and we had a lovely afternoon at Coate Water.

The next year we went further afield to a farm near Petersfield. We were expecting about thirty teenagers and the only way to get sufficient cake for the youngsters was to make it ourselves. For three Friday evenings before we went, the girls came to our house and we spent the evening cooking. We had the Rayburn to use, plus my electric cooker and we found the calor gas stove we used at Barn Camps and put that in one of the bedrooms. Everyone got stuck in and we churned out cakes by the dozen. The kitchen temperature rose by several degrees as the seven or eight girls and I cooked for two hours. I then had to bag everything up before going to bed and as we had a large freezer was able to store the cakes there until camp. We did this every year that we ran barn camps and the cakes were a great success, I usually made apple pies as well.

We had a Land Rover by now; it wasn't new and had no seats in the back but we managed to squash eight teenagers in. Gordon also had a Land Rover so squashed another eight in it. We borrowed a small caravan for the children and me and had two barns, one for each sex. We had bought a second hand Calor gas cooker at an auction sale and Dave had managed to add an extra shelf. Someone had given us a small marquee and we had bought some second hand chemical toilets. These were rigged up behind some hessian and seemed to be quite adequate. I asked Peter when we got home how he had managed as he was still quite small, "It was all right Mum, I held on to the tree!" That year we had no tables from which to eat – or chairs- so had to balance food on our laps, the only table we had was for the food preparation. We had two brothers with us, Thomas and John who had been going to the Methodist Sunday School. Their leader asked if they could join our Bible Class as they were much older than the other children in the Sunday school so they came to camp too. On the Saturday everyone went to Petersfield, there was quite a fleet of cars including two Land-rovers, which each held six in the back and three in the front.

They came home at six o'clock full of their day out, I had a meal ready for them and we were halfway through it when somebody asked, "Where's Johnny?" Nobody had seen him since he was last spotted in a boat in the middle of Petersfield pond! Dave drove the ten miles back to Petersfield and decided to start looking in the Police station. Johnny was there. He had no idea where we were camping so the police thought we would discover he was missing sooner or later. He had a habit of getting lost. One year we took the teenagers up to London for a sightseeing day. John's absence was discovered when they were looking around the Tower of London. He was found in a Beefeaters hut. "Have you lost John Fisher" they asked. He had even had a cup of tea while there. Another boy was lost one year in London on a snowy December day. He said afterwards "I bent down to pick up a sixpence and when I got up you had all vanished." Dave had no idea where he was, so went to Scotland Yard, lining up all the others outside. Mark was found back at the cars. "Well, you said that if we got lost to go back to the car."

The highest number that we ever took to barn camp was eighty as other church groups joined us. By then we had bought more efficient cookers, trestle tables and other equipment. We were staying on a farm near Winchester so had decided to spend Saturday afternoon in the city and have a 'Spot the Officer' competition. This consists of the helpers disguising themselves and walking round the town while the teenagers tried to spot them; when found there was a password to exchange. At a certain pre-arranged time the challenge was over and whoever found the most officers, declared the winner. That particular year one chap shaved off his beard and turned himself into an American tourist with dark glasses to complete the disguise. John became a market researcher gathering signatures to prevent the removal of King Alfred's statue to America. The Queen Mary had just been bought by America so it was quite topical. I think it was a different year when David borrowed the vicar's 'dog collar' and clerical vest and sprinkled talcum powder in his hair to turn it grey. He spent the afternoon in the church porch reading the notices!

We usually had a guest speaker to give the epilogues each evening. We would have a good sing first accompanied by guitars played by various people and then the speaker spoke for twenty minutes or so. Once again we would listen to God's Word and hear how the Lord Jesus came to earth to save us from our sins if we trust in Him. We had one man come to speak who was excellent but after the meeting was quite despondent. "How can we stop those birds from making so much noise?" The pigeons were roosting in the rafters above our heads. What Colin didn't realize was that it was an everyday noise to most of us and we, and the youngsters hadn't even noticed it; he lived in Southampton.

One year when 'hot pants' were all the rage, I had altered jeans for David and dungarees for the boys. I had manufactured shorts with a bib and cross over straps and even managed to find a teddy bear to put on the front of Dave's. We were coming home along the A4 on a Bank Holiday, towing a caravan when we had a puncture in the back car tyre. Dave pulled into a lay by; I had no idea where Dave's trousers were so he had to change the wheel wearing his teddy bear adorned hot pants! Another year we had a theme of cowboys and Indians so I made waistcoats for the three of them, they already had checked shirts and I made kerchiefs for their necks and bought cowboy hats in the toy shop. Every evening before we began the evening get together, Dave would sing 'Camp Diary'. This was an account of things that had happened during the day, which David and I made into a song with a catchy chorus, which everyone joined in with gusto. 'Home, home on the Range' was the tune one year.

Some of the boys played their guitars each evening and during the evening others would take part in the programme. One year the guest speaker was also a conjuror and kept us all spellbound with his tricks. He stood up on the table and produced 'watches' from thin air. He also had a fund of silly stories with sillier endings such as the man in the burning building. He built the atmosphere up with the description. 'Then the man went up to the next floor and poked his head out of the window. We shouted "Jump, man, jump into this tarpaulin we're holding for you!" Would he jump? Would 'e 'eck!' In the end, the man was on the top floor. "Come on, man, jump into this tarpaulin we're holding for you." So he jumped – and missed!'

After about half an hour of stories and silly songs everyone settled down for the more serious time when John or Colin or Phil would once again tell us of God's love that sent Jesus to this world to die for each one of us; that we might know the forgiveness of our sins. We would have a mug of cocoa before we went to bed but there were often times that really made these youngsters think about the way their lives were going.

23. Furniture Auction

THE HOUSE WE NOW LIVED IN was one of a row of five and quite big for a cottage as it was two made into one so we had two large bedrooms and two small ones. Down stairs was a utility room, kitchen, bathroom, one big sitting room and a living room and we badly needed some more furniture. Somebody suggested that the Naafi sales at Ludgershall were good sources of good second-hand furniture so I decided to go and try my luck there. There was so much stuff that the sale was spread over two days as there was not only furniture for sale but curtains, crockery, every kitchen appliance one could think of, some clothing, which was new, the list was endless.

Dave came over on the viewing day, which was held on the day prior to the sale and marked in the catalogue the things that he wanted and the price that he was prepared to spend. The next day Andrew and I drove the ten miles and at ten o'clock a man went round ringing a big handbell, Andrew who was about two years old was mesmerized. I had him in his push chair and hoped that he'd stay put, which he did.

That first day I bought two settees and three arm chairs at knock down prices, £1 for each of the settees and £5 for the three armchairs. I also bought a quantity of material to make chair covers and curtains for about £5. Once I got into the swing of things there was no stopping me. The furniture had come from Naafi hotels and was good solid stuff and hadn't been used much. I learned that the way to bid was to wait until somebody else opened the bidding and not to jump in first. I bought three pairs of golfing shoes, brand new – with studs in the soles! Well, I thought for wearing on the farm they would be ideal, the smallest size was size six but I reasoned that although I only took fives, with a pair of thick socks they would be ideal. The other pairs I bought for friends; I felt sure that at £1 per pair they were good value!

I was back on the second morning, the sale started at 9 a.m. so I had quite a rush to get there; the first job was to find somewhere to park as the sale site was surrounded with residential roads. Andrew thought it was great fun when we had to run to get there in time.

I found a good place to stand, gave Andrew some toys with which to play and once again we were off.

"What am I bid for this fine pair of vases?" the auctioneer started. They were one thing that I didn't want! Hideous bright orange with dragons for handles.

David wanted a filing cabinet.

"A fine filing cabinet. Who'll start me off? Ten pounds? Well seven fifty then? Somebody start me off."

There was complete silence. I gathered up my courage but someone was before me.

"Three pounds."

"Three pounds fifty."

It sold for five pounds. There were some other cabinets so I was ready this time

"Four drawer filing cabinet. Come on, someone start me off..."

"A pound!" My voice wavered.

"Madam, I need a sensible bid."

There was complete silence.

"Come on now, or I'll have to take this bid, ridiculous as it is!"

Nobody spoke. The gavel fell and I was now the proud possessor of a filing cabinet. The next one went for £5!

There was some attractive lengths of new material two toning lengths were in each lot just right for curtains and a tablecloth, with enough material for six table napkins. I actually still have them forty five years on.

One year I bought two stainless steel sinks complete with draining boards, which we used at camp for some years.

Over the years I returned to Ludgershall Naafi sales and was sorry when they ceased. I went to other sales but none had the same atmosphere as they did.

A pub was selling up to make room for development in Andover and Dave and I went to the viewing day as Dave wanted a work bench and thought there might be something suitable. There was a stained bar table, quite solid, and just what he wanted. He couldn't come to the sale so once again I was given a target price of ten pounds.

The bar chairs and small tables were sold; I bought a bundle of snooker cues for a friend, then bought two chicken runs and housing for £1 each that Dave had thought another friend wanted. I was on a roll!

"What am I offered for this refectory table? Who'll start me off?"

I joined in and in the end there was two of us; I was standing next to the auctioneers clerk and happened to see that the bidding was at £150!

"I'm sorry," I stammered, "I thought it was £1.50!"

"Madam, please make sure you know what you are doing."

I crept away and hid myself!

Eileen in cut down brownie uniform.

Joyce Pam and Eileen, ready for church.

At Aldershot Bathing pool.

David with sisters Maureen and Patricia plus parents.

Cast of 'Toad of Toad Hall. Eileen kneeling, fourth from right.

On a Bible Class ramble. David and Eileen, early days.

Off to camp on a coal lorry.

Off to camp by train. Eileen extreme right.

Our Wedding, 1961.

Our first house.

Billy Graham, who made a big impression on us.

Son Andrew in 'standing' box.

Bathtime for sheepdog Briar.

Our present house before alteration.

And as it is now.

Granddaughter Bethany feeding a lamb.

Granddaughter Andrea feeding Paddy.

Miriam helping herself to dog biscuits.

Sammy and Paddy.

Trying out the new dog kennel.

'Sheep' transporter.

The Sullivan Family.

24. Growing Boys

WHEN ANDREW WAS SEVEN, he had to have his tonsils out. I had had mine out the previous year, not a very nice experience but it was good to be free of a continual sore throat. I was kept in for six days as my tonsils were difficult to remove, they had shrivelled. Andrew came home after five days in Odstock Hospital and there were only a few days before the end of the term. He loathed it in hospital – mainly because they served cooked teas and he wanted bread and marmite! I was back in Odstock a year later to have two impacted wisdom teeth out as I was still getting sore throats and it was felt that the wisdom teeth were to blame.

A boarding school was closing down and selling all the fixtures and fittings. I obtained a catalogue and we discovered that some cricketing gear was being sold plus some desks among other things. Andrew wasn't at school because he was recovering from his tonsil operation, so I decided we would go to the auction. We managed to get the cricketing gear although the pads were much too big but there was a set of stumps and a cricket bat. There were three old fashioned school desks, one being sold singly and two as a pair, the single one was being sold first, which I bought for two pounds and later was able to sell it to a neighbour for her little girl. The two in which I was interested came next. They were obviously quite old with hundreds of names carved into the lids.

"What am I bid for this pair of fine Victorian desks?"

My heart sank. If they were Victorian they would probably go for more than I wanted to pay for them.

"Does no one want them?"

"Two pounds," I quavered.

"Don't be silly Madam. They're genuine Victorian."

There was dead silence.

"All right Madam, they're yours!"

Dave's father renovated them, turned one lid over so that the carvings were underneath and replaced the other lid. The boys used them throughout their schooldays and I sold one for forty pounds and the other for twenty pounds some fifteen years later; it would have fetched more if the lid hadn't been turned over. I also bought a gross (144) squared red exercise books for a pound, they lasted us for years for various uses.

The farm gardener and his wife lived next door and she'd look after Peter when I did my shopping while Andrew would spend the couple of hours 'helping' Vera, a friend , to collect eggs from the chicken battery that was on the farm where her husband, Gordon was manager. They also worshipped at the Gospel Hall at Vernham Dean where we now went to Church.

As Andrew grew older I needed to find a play group for him but there wasn't one in the village so I had to take him the two miles to Stoke where there was a little nursery school. I'd leave him on the two mornings that he attended, protesting at being left but the staff assured me that he soon stopped crying once I had gone. It was vastly different from the Play Groups of today, with a lot more structure; they were even taught writing although Andrew was no good at it and even today his writing is not too special. The only other thing I remember about it was at the end of term they put on a 'performance', this consisted of the children dressing up in various costumes – Andrew was dressed as a jockey – and had a rhyme to say as to why he wanted to be a jockey; he ended up as a farmer, though not with Dave.

By the time Peter was old enough to start at playgroup one had been opened in the village. There were quite a few pre-school children and it started off with eight children attending; it cost half a crown per session! The mums had to help on a rota system, which was nice because we got to know other people and their children, which prompted cups of tea in the afternoons and swapping children.

25. A New Bungalow

AFTER WE HAD BEEN IN HURSTBOURNE for four years, David's employers decided to have a bungalow built for us; we were able to have some input as to the design and found it was nice having a new house instead of the old ones we had lived in up to then. It had four bedrooms, one of which Dave used as an office. The boys shared a room and I was able to keep one as a spare as we had a lot of visitors. We also had a large kitchen diner and a lounge, plus a garage and big porch where the boys liked playing when it was wet.

The bungalow was the other side of the field behind the house where we lived so when it came time for moving in, Dave mobilized the Youth Club he was running at the Gospel Hall to help us move. Andrew's bed was carried over the heads of two of the boys; it was raining and they wanted to keep dry! Eventually everything was in the bungalow and it didn't take long over the next few days to settle in.

Andrew was at school by then. When I took him down to see the school before he started I was put off by his teacher when I asked what she taught the children. "Oh, I don't teach them a lot as they're only village children." I had to take him down in the morning break time as I was told it would be too intrusive if it was during a lesson. My boys were not the only ones who went on to University but it seemed a very disconcerting thing to hear.

Both boys started to learn to play the trombone while they were at the village school; a peripatetic music teacher went to the various schools for half a day each week to teach brass instruments. Peter turned out to be very musical and they both enjoyed their brass lessons. When they started at the secondary school, the master also taught brass there but Peter gave up the trombone and started learning the tuba. Nobody wanted to play it and Peter saw it as a challenge – so did I when I had to bring it home one week to try and polish it up, the coach driver wouldn't have it on the bus so I had to drive into Andover and fetch it! At the secondary school concert every year Peter used to play a solo and brought the house down when he played 'Tuba Smarties' by Herbie Flowers of the group that was popular then called 'Sky.' He rigged up fairy lights all over his tuba and certainly made people realize that music could be fun.

Another year he played 'The Elephant' from 'The Carnival of the Animals!' Andrew still played the trombone and was in the school orchestra with Peter. Peter also had piano lessons as we had been given a piano and he proved to be able to learn how to play. He still loves music and is responsible for the music group at the Church he attends in Aldershot. It's the same church we attended when we lived there before we were married.

David goes there to preach sometimes and it's such a thrill to meet up with friends I knew in my teen age years and also the people that helped us in our Christian life.

David was well established in his employment; the sheep flock had improved and his employers bought some land a few miles away. This was quite a challenge as it had been neglected and needed a lot of TLC. The tractor driver's wife and I did quite a lot together as she had three children a bit younger than my two and the five children got on well together. At harvest time they loved to be able to take tea up to the men who were able to stagger their work and eat their tea without stopping the combine. As the men finished much later than the rest of the year, tea time was the only time that they could see the children. It was usually about ten o'clock before they were finished and sometimes Dave would be later than that if the dryer needed attention. Tea time was also an opportunity to pick blackberries, which grew in profusion around the fields; I'd put the children to bed and then set about making jam.

We loved our holiday times and found that the nicest way to have one was in Dave's Dad's small caravan, which we would hitch up behind the car and just take off. When the boys were older we would take their tents with us and they would sleep in those; we would eat more baked beans in that fortnight than we did the rest of the year. We enjoyed going to Wales and climbed Snowdon by two separate routes the Watkin path was difficult but Briar, the dog loved his long walk

He was lovely to take away with us and sat on the back seat between the boys, he could see out of the front window, but had the annoying habit when he was young of chewing everything he could including a jersey and the cable from the caravan to the car. The last family holiday we had was to Scotland, we drove up to Inverness where we collected the cabin cruiser we had hired, and cruised on the Caledonian Canal for a week travelling from Inverness to Fort William. We stopped and looked at the 'Nessie' exhibition near loch Ness but never had a glimpse of her. The locks on the canal were big affairs holding a dozen or more boats; they were operated by lock keepers who stopped for lunch for two hours and knocked off at five o'clock! The villages we stopped at were usually nice and we could get stocked up on provisions. Andrew and David went up Ben Nevis, which, after Snowdon, seemed like a stroll in the country and having equipped themselves with walking boots, provisions etc. they found it was easy. Even a man in a wheel chair was being pushed up the path. Peter and I had a lazy day on the boat; Peter wanted to get his applications written to secure a university place the following autumn and I caught up with writing some post cards.

We decided to go to church on the Sunday and found one that read "Free Church." It was like no other Service we had been to before or since. Three 'elders' sat in chairs at the front facing the congregation; there wasn't a sound before the service only the rustle of sweet papers from the young children who were there. We stood to sing hymns, pray and read the Bible, everyone seemed to be dressed in black and as soon as the service was over everyone disappeared almost before the final 'Amen' was uttered.

We had a couple of holidays on the Norfolk Broads, which were totally different from either the Caledonian Canal or the Oxford Union canal along which we travelled on a narrow boat one year. The cabin cruisers were much faster and there seemed a lot more boats about. We had our first taste of stock car racing when we were in great Yarmouth; all I can remember is the noise and the smell! Whenever we stopped, Briar, whom we had with us, would jump off the boat and go exploring; unfortunately, he'd sometimes come back, looking very pleased with himself, smelling absolutely foul having found a dead animal or other rubbish in which to roll. He did swim but wasn't keen on it so we would banish him to the other end of the boat.

One year we had arranged for a holiday in Wales. At nearly the last minute Dave decided that the car needed a new engine but felt that a reconstructed one would be all right. The day before we were due to leave the engine was delivered in four cardboard boxes! He spent the next day working on it and we finally left – minus the dipstick –at seven in the evening! We made it to Monmouth where we were staying the night at a friend's farm and continued our journey to St. David's, which has a lovely cathedral and, so, we were told qualifies it to be the smallest city in the British Isles. The scenery is lovely there and the beaches big and sandy. We found a nice caravan park overlooking the sea, which Briar loved but couldn't understand the waves and kept barking at them. One year when we were staying in the area we discovered a T.V. company was filming the C.S. Lewis classic, "The Voyage of the Dawn Treader," from the Narnia series we were fascinated by the perfection that had to be achieved.

Another year after the boys had left home we went back to St. David's by ourselves and stayed on a farm. We went to a performance of "Othello", which was being performed in the ruins of the Bishop's Palace. It was pure magic. We had our rug to sit on and as the floodlights came on it was unforgettable as the actors used the ruins to bring the play to life. We also went to a 'Welsh Evening', which was an amateur production but the various items were well performed.

26.   A New Farm

AFTER THIRTEEN YEARS AT PARSONAGE FARM, David was offered a job by another local farmer who needed a manager. This was a smaller farm but had sheep and arable; David would be able to keep some sheep of his own, which he had wanted to do.

Our next house was thatched and had to be extended, so, for the first year we lived in the adjoining one, which was *very* small but a door was temporarily made through into the other house so we had the use of the lounge next door. The previous tenants had had two children, both girls so as the boys 'inherited' their bedroom they had to put up with ballerinas on the wall paper. We knew it would just be a temporary move as we would be moving into the house next door, which Mr Trewby was having extended for us. Once the stairs had been altered Andrew moved into a bedroom in the other house it was partly used for storing our extra furniture so he didn't have a lot of space. We had a small kitchen, living room, hallway where Dave kept his large roll top desk, with two bedrooms and a bathroom upstairs. When the extension was finished we had the large lounge, big kitchen, dining room utility room and study, with four bedrooms, box room and bathroom upstairs.

Dave had twenty five sheep of his own, which ran with the farm sheep. The farm was smaller than his previous one but had plenty of challenges to cope with; the farm had been making a loss so Dave was determined to alter that state of affairs. There were other animals kept, a few pigs, Daisy the house cow who had to be milked twice a day, lots of chickens running around, and a couple of ponies – it was a typical village farm. There were some fields around the farm but some of the land was four miles away, this was where most of the corn was grown. At harvest time this had to be hauled to the grain dryer for drying, which was by a track in a different part of the farm.

There were a lot of outhouses and farm buildings around the farm and the boys had great fun exploring as the buildings were full of interesting objects. In one shed they found about twenty lawn mowers! Mr Trewby hated throwing anything away as he said that it might come in useful one day – and it usually did.

Lambing here was supervised at night as well as during the day; Paul, one of the other employees helped with this as well as Sally, the farm secretary. Last thing at night was always a bit hectic in the kitchen with bottles being prepared, instructions being given and coffee always available. There would be the odd lamb either coming in to be warmed up in the oven or one just vacating it ready to be returned to its mother. Eventually we would get to bed, Dave having been on duty since five in the morning.

It didn't seem long before we were woken up with a tapping noise at the window. Dave shot out of bed and opened the window. Paul put the clothes prop down, which he had used on the window to rouse Dave. "Dave, can you come, there's a ewe with a lamb stuck, head and no legs!" Dave knew what he had to do but it was a tricky procedure as the head would have to be gently pushed back into the womb then the front legs extended and the head and front legs pulled out together. He has very large hands, which help a lot but which can't be very comfortable for the ewe.

Dave hastily pulled on his trousers and jersey over his pyjamas, "Won't be long dear."

I turned over and went back to sleep and slept until David awoke me with a cup of tea. He's always maintained that if he had to get up in the night he'd roll me over to his side when he went out and roll me back when he returned so had a warm bed.

27. Students

ANDREW OWNED TWO FERRETS as he and a friend found one when they were rabbiting one day; Jill was a typical sandy coloured polecat ferret. Toby was Andrews other ferret, very tame and a dirty white colour; Andrew brought him indoors sometimes and he soon learnt where the cat's food was and would help him to clear it up. Andrew used to take him for walks along the lane having bought a harness for him to wear. He needed a new cage, as because he kept chewing his way to freedom, a wooden one wasn't ideal. Scouring the local paper one day we found notice of a sale at the winter quarters of a circus so decided to go to it. There was a metal cage, which we felt would be ideal for Toby so we decided to try bidding for it and were very pleased when we were able to buy it for £5. Andrew had sourced dead chicks for sale so we would buy 5lbs at a time, which he'd divide and bag up in lots that would last him for a week, and then they would go into the freezer for Toby and Jill; we always seemed to have dead chicks thawing out for them to eat.

Yew Tree Farm was smaller than Dave's previous one but the boss had expanded it over the years and David found it quite challenging. There was a lot of arable land apart from the sheep.

When we had been there a couple of years David was approached by Sparsholt Agricultural College to ask if he would be willing to teach practical shepherding tasks to a new class that was starting that Autumn. This would entail one day each week either at the college or on a designated farm. Would he also be willing to take a student one day every week; other farmers were also being asked to take a student. Dave had to talk it over with Mr Trewby who was quite agreeable and they were able to come to an agreement about the day off each week. Dave found this new venture very challenging. So did I! During lambing we had our designated student or two living with us for a fortnight so I always seemed to be cooking meals as well as reviving baby lambs in the bottom oven of the Rayburn. This was usually in March with inevitable wet weather so there was always wet clothing steaming on the Rayburn rail and soggy students needing dry clothes. The baby lambs needed an endless supply of old towels in which to wrap them for their bottom oven incubator. Towels were also appreciated by the students to stop the drips going down the backs of their necks.

I found that hand cream was appreciated as although plastic gloves were all right for some jobs, hands had more sensitivity. I gave up washing the kitchen floor as it was usually where revived lambs took their first stumbling steps and welly boots had a corner of their own. The girls we had over the years seemed to have more of a 'feel' for the lambs than the boys and we enjoyed their company. David always did the early morning shift as he's never had any problems with getting up in the morning, he was usually out in the field by 9a.m., in spite of having been very late to bed the night before. Over the years we had countless students with us and are still in touch with several.

Jane wasn't one of the Sparsholt students but her parents moved to the village from London where her father had a building business. He was coming up to retiring age and they had decided to buy a house in the village. We had a lambing open day one year when we invited people to come and see the sheep; Alan and Daphne came to that saying how much Jane would have enjoyed it. Jane went to Cirencester Agricultural College where she met her husband and eventually her parents bought a farm in the next village where she and Mark farm with their three sons. Our son, Andrew has worked there for some years now although he lives ten miles away at Tidworth.

28. Sheep Dogs

WE USUALLY HAD DOGS RATHER THAN BITCHES but Dave decided that he wanted to breed his own replacements so Jan was purchased. In due time she produced five puppies, one of which had a deformed jaw so had to be put down. Dave sold three of the others and kept Jill for himself. She was an amiable dog, always wanting to please and Dave was pleased with the way she worked. I was at home one day with Jill as Dave was doing a job for which he didn't need a dog. I was busy indoors and suddenly heard a commotion outside. I rushed out to discover that a neighbour's dog had hold of Jill and was shaking her like a rat. I ran to find the owner who came running out and managed to drag the dog off. Jill was covered in blood, especially one front paw. There were no mobile phones then so I lifted her into the car and drove to the vet.

"You'd better leave her with me; we'll do what we can for her. Tell David to phone this evening."

I drove home wondering what Dave would say. He was, understandably upset but, as always, kept calm about the situation. We collected Jill the next day. One of the pads on her feet had been half ripped off, which was the worst injury, but she had several other nasty lacerations. She was equipped with a 'boot' to wear on her foot and an 'Elizabethan' collar to wear round her neck to stop her from trying to pull the bandages off. She was terribly bored at home and I had to take her for walks. One day she went missing and by the time I found her she had nearly got the boot off and had bitten the bandages, so she had to go back to the vet to be rebandaged.

The dogs were always frightened of loud bangs, thunder, and fireworks! We always tried to secure them in their kennels on November 5th, but now there seems to be more excuses for letting fireworks off on other occasions. One year, Bud, one of our other dogs, chewed his way out of his kennel, but Dave managed to find him. We've always had trouble with dog tags as with a tag they tend to get ripped off and collars don't have a metal piece any more. Bud is micro chipped but unless a dog is found by someone prepared to take the dog to a vet to have it read, it is not a lot of help. One evening we had been to the Children's club which we run and on stopping at a house to deliver a child were asked if we had lost a dog as somebody had called in to say they had just found one. We lived the opposite end of the village and felt sure that it wasn't any of ours but when we arrived home, we discovered that Bud was missing. We made a phone call to the child's mother who had the telephone number of the people that had found the dog – they lived at Hungerford, ten miles away; needless to say it was Bud, his kennel now resembles Fort Knox!

Living in a thatched cottage had its drawbacks we discovered. I had lit the fire one afternoon when I realized that the chimney was on fire! I dialled 999 and on explaining why, was told the fire engine would be with me shortly. I grabbed the cat, went outside and could see clouds of smoke coming from the chimney. This was before the days of mobile phones so I couldn't get hold of Dave. At last I could hear the fire engine not one fire engine but two arrived; they explained that if a thatched property was involved, two appliances were always sent. A few minutes after they drew up, Lady Betty Cuthbert arrived. She lived further round the lane in a Queen Anne house to which they had moved when her husband, Admiral Sir John Cuthbert retired from the Navy.

"I saw the fire engines and wondered where they were going, Eileen. I was chief firewoman in London during the war so am always interested in fires."

She greeted the fire men who soon had the fire under control. They all knew her. "Now Eileen, put the kettle on as I expect the men would love a cup of tea."

David was surprised to see all the water when he came home but got up early the next morning to sweep the chimney. He was on top of the roof with a torch as it was still dark; he quickly changed its position, as it was shining into the uncurtained bedroom window of Mary, an elderly neighbour. The boss, Mr Trewby had a wood burning stove put in, which was much safer than the open fire.

29. Boom Years

VARIOUS SEASONAL WORKERS ARRIVED in August to help with the harvest, mainly New Zealanders. Doug was with us for quite a while living in a caravan but used our facilities and had a meal with us each day. He always wore flip flops on his feet no matter what job he was doing – even driving the combine harvester. He had an old Renault van and when he finally left to go on to America for their harvest, he told the boys that they could have it. The boys were overjoyed. Andrew was about fifteen and Peter two years younger. They took out all the windows , painted it, fixed cow horns on the front and once they had pushed it over the road could travel for miles on cart tracks. 'The Dukes of Hazzard' was on the television at that time so they had someone to copy. I think that learning to drive like that helped them when they were old enough for a driving licence, Andrew passed at his second attempt and Peter at his first.

They've both been careful drivers although Peter had one accident when coming home late one evening and wrote his car off. He hit a bank and the car ricocheted off the bank on the other side of the road and demolished a line of fence posts. He replaced them for the farmer and had to travel some other way to get to college; both boys had motor bikes so he used that. Dave still has an old one too, which he sometimes uses on the farm, all three of them had had motorbike lessons, Dave enjoyed his lessons, which he had with half a dozen teenagers.

In 1980 somebody suggested to Dave that he apply for a Nuffield Farming Scholarship as they were only available up to the age of forty. As I have written at length about it in "The Sheep's in the Meadow-Hopefully"! I won't go into it again. I found it very hard at home without him, although the boys were there. They were both at Sixth Form College but were studying for 'mocks' so were very busy. I picked up a 'bug' and was very unwell over the Christmas, which we spent at Somerton. I stayed with Dave's parents and the boys stayed with his sister. We phoned Dave up on Christmas morning, which made me feel worse. He was in hot, brilliant sunshine and we had cold, damp, foggy weather. He said that it felt strange at a church service he attended to be singing 'In the bleak mid-winter' dressed in shorts and a summer shirt. Andrew had passed his driving test and although he hadn't done much driving managed the trip to Somerton well.

Both boys went to Sixth Form College after leaving school. Andrew needed to do some retakes of his GCSE's and a couple of 'A' levels before securing a place at Seal Hayne Agricultural College when he was eighteen. Peter wasn't sure what he wanted to do so took A-level Maths, Physics and Electronic Systems.. He went to Surrey University at Guildford where he studied Electronic Engineering, which was what computer studies was all about and became a B-Eng.Hons. Andrew was based in Devon on the edge of Dartmoor and couldn't get home very often. Peter was very involved with various events at Uni but came home when he could. He belonged to the Gilbert and Sullivan Society and we all went to see him perform in The Mikado. We had a meal first in the student house in which he lived and then all six of us – the boys and both their girlfriends – crammed into our car to go the mile or so to the University. It was a lovely evening although some of the songs, which had been adapted for the performance were un-understandable to non-students.

His graduation ceremony was held in Guildford Cathedral, which was next door to the University. By this time he was engaged to Brenda so we had her parents from Cumbria staying with us for a week. They had come down by train so came with us in the car. It was such a thrill when the Duke of Kent shook Peter's hand and presented him with his degree.

David had had an offer of more part time work at Sparsholt. It was decided that the lowland sheep course at the college, needed more practical input and David was asked if he'd teach practical shepherding skills. As well as that, each student was allocated to a farm in the area where they would spend one day each fortnight. So one week Dave would be teaching one group of ten at the College and the other ten on the following week. This arrangement gave him a fresh interest. He was very good at teaching the practical skills with the highlight of the year being sheep shearing. He had to arrange with the various farmers when and if the students could shear their sheep. Some became very proficient at it and David is still in touch with some of his students some thirty five years later.

The shearing had to be organized like a military operation with eight students at a time. They would turn up at 9 a.m. before which the sheep would have been rounded up and brought to the barn where the shearing was going to be done. Dave would sort the students out so that one would be keeping the catching pens topped up with sheep, a couple would be shearing, another two were responsible in rolling the fleeces, and one sweeping up the daggings and so everyone had their part to play. The sights, sounds and smells of the shearing shed are unforgettable with everyone playing their different role.

Dave would patiently stand over a student showing him – or her- just how to hold the sheep and shearing handpiece. The greasy smell of sheep and oil as the handpieces were oiled continually and the non-stop bleating of the sheep as they milled around awaiting their turn, is memorable. My job amid all this cacophony was to make sure that the students were fed properly with buns, biscuits and coffee mid-morning, a more substantial sandwich meal at lunchtime and more cakes at afternoon 'smoko'. Everything was eaten, which made it gratifying.

At lambing time each student spent time at their adopted farm helping out and learning as much as they could. It was another physically demanding job with no set hours as sheep don't watch the clock. The weather was not always conducive but with so much lambing being done indoors now, it didn't always matter so much. The first lamb that a student delivers was something special and a real sense of achievement but as there were still five hundred ewes to lamb, there was no time for reflection. A lot could go wrong and usually did, especially when it was raining.

They were allocated three weeks over Easter and so had to sleep at their adoptive farm as well. They were usually cheerful and appreciative and it was a unique way of finding out what shepherding is all about. When it's blowing a gale in the middle of the night shepherding isn't about white fluffy lambs but soggy little rats that are suicidal. Lambs usually die of one of three things; – drowning; the smallest puddle and they lay down in it having escaped from what was thought to be an impregnable pen. Strangling is another cause of suicide; they put their heads through the most minute noose and walk round and round until they die. Their mums also have a habit of laying on them so they suffocate; they then search for it bleating piteously, "Where's my baby?"

The sheep utilized grazing on farms other than their own, which necessitated a lot of sheep movements along the roads. Dave had a lovely girl on the shearing course who lived locally so would ask her and her friend, the farm secretary to help. Jane would walk in front with her crook and Sally usually walked at the back. Both girls would wear shorts when the weather was warm enough. Dave drove in his van ready to pick up any stragglers. He liked to move them early on Sunday mornings when the roads were quieter but it wasn't always possible. He found that if Sally walked in front with Jane bringing up the rear drivers were more considerate, he'd drive a couple of cars behind in case of any difficulties. They were moving one flock of sheep along a country road when a police car came towards them. On seeing the sheep and an attractive young lady leading them they promptly turned their car round and drove in front with their blue light flashing.

At one time Dave was utilizing land on 34 different farms and the paperwork was quite incredible as there were about 5,000 sheep on keep. The sheep would come down from The North at nine months old and Dave would rear them for the various owners. The furthest that sheep could comfortably walk in a day was about four miles so if David worked out that it would take longer, he'd have to arrange for an overnight stop on the way. It wasn't easy to find grazing for three hundred ewes, the most that it was practical to move at one time, but what made it possible was the invention of the Ridley Rappa electric sheep fencing.

This could be erected very quickly and he could just put a fence round the corner of a field and the sheep could spend the night there if it was a long journey. The Ridley Rappa system was a system invented by Harry Ridley, which became a tool that made it possible for David's scheme of 'gipsy shepherding.' It consisted of three strands of wire, which are electrified, with anchor posts in between. This was connected up to a fencer unit and battery and quickly erected wherever David wanted to stop. A gadget was eventually marketed called the Ridley Barrow, which was a lot quicker than doing it manually. When Andrew finished at college he and Dave started up another business supplying and fitting Ridley Rappa fencing in the area.

There was a farm five miles away that had a big outdoor pig farm and Andrew had the job of moving miles of fencing every few weeks. We had a nephew who had just finished at university and wanted a job while he looked around for THE job. Clifford, our nephew, came to live with us for a few months and found the discipline of erecting the Ridley Rappa fences very helpful. He went on to manage an otter project in the St Ivel Valley near Biggleswade but has always said how helpful he found the time spent with us.

The 1970's and 80's were the boom years for sheep farming. Andrew was able to go to New Zealand for six months as part of his pre-college work experience and learnt lot about life as well as sheep. Steve, a friend of Andrew went with him and they were able to get jobs on farms through Dave's Nuffield farming contacts. The farm on which Andrew found himself grew trees as well as rearing sheep and part of his job was thinning the lower branches from the trees. They were both able to go for Christmas to stay with our friends Nigel and Marion, with whom David had stayed the year before. They also spent a fortnight touring South Island in an old car that they bought cheap and met up with many characters on the way, camping out most nights. They returned to the U.K. via Los Angeles but by this time were almost out of money. They stayed in a bed and breakfast place and lived on doughnuts as they were so cheap. They found the cheapest thing to do was to be part of the audience at the many T.V. shows that were on. They had an unexpected extra night in L.A. due to a fault with the plane so were put up at a good hotel. Andrew managed to find a job on a local dairy farm then went back to Seal Hayne for his final year.

30. Share Farming

DAVE NOW OWNED SEVERAL VEHICLES; a Land Rover, motorbike, sheep transporter, a stock trailer and a second hand tractor. He found the motor bike was most useful on the downs as he could follow the tracks made by the sheep. At one time he had sheep in nine different locations so needed to be able to move quickly between them. He had a wide wooden tray to fit on the back of the motor bike so he could carry a dog. He was once filmed for 'Farming Today', which included a shot of Dave on the motor bike with the dog on the back Eventually, Dave realized that to carry on he would need to be self-employed but as we were living in a 'tied cottage', this created problems. He had been working part time for a year and then Mr Trewby's son wanted to come back and manage the farm so Dave gave in his notice and we started house hunting, which wasn't easy as we wanted to stay in the same area and there was nowhere to be had.

Peter by now was at Surrey University reading Electronic Engineering but still needed a base and Andrew was coming to the end of his three years at Seal Hayne Agricultural College. He was engaged to be married to Rosemarie a farmer's daughter from Devon but they met when she was working as a veterinary nurse in Marlborough, and Rosemarie started coming to the Gospel Hall.

David then worked out a system for Share Farming sheep. It had been done with cattle but nobody had been successful with sheep. It all started when David had been able to graze his small flock of sheep on some land belonging to Mr Gerald Boord. Mr

Boord was concerned that by David using his land it might be construed as the occupancy creating a tenancy and he would not have control of the land. David worked out a scheme that by Mr Boord owning a percentage of the sheep and paying Dave to shepherd them it would get round the problem. Any profit made by selling the ewes or their lambs would be dealt with in proportion.

This system worked well. A friend of Mr Boord was in a similar situation of having land but no livestock so he entered into an agreement with David. Then others wanted to join in, some of whom owned no land so they rented land on which sheep could be kept. In the end there were eight farmers in the syndicate and David realized that he was no longer able to be employed, so took the step of resigning from his job to become self-employed. The sheep were year old lambs and were kept for a year when they were sold for breeding.

Dave was now shepherding sheep in three counties: Hampshire, Berkshire and Wiltshire. He used one field where the three counties joined. He had a van to begin with but eventually bought a second hand Land Rover. He also had his motorbike; he adapted his trailer so that he could carry the motorbike on it; when he reached the location where he was going, he could use the bike, which was easier than the Land Rover.

We now needed somewhere to live in the locality, as we were in a tied cottage.

31.   A New Cottage

EVENTUALLY WE HEARD OF A ONE BEDROOMED COTTAGE which had become empty, the tenant having died of old age. We went to look at it; it was just where we wanted it but it would be small even for a couple, with no room for the boys. We looked at the problem from all angles; it was a Grade 2 listed building, which could make things difficult, as we wouldn't be able to alter it much. We drew plans up – and scrapped them. We consulted friends. Then we eventually worked out that if we could build a modest extension we would have room for a kitchen and utility room. The present little dining room was just big enough for a bedroom and the large landing would make another one upstairs. Of course all this would need planning permission and a co-operative bank manager. In the mean time we needed somewhere to live and found a house to rent in a nearby village on a short term tenancy. At first our planning application was refused but with a bit of juggling we got the go ahead.

The village in which we were staying temporarily was a bit out of the way but it was a comparatively new house with plenty of room. It was a very cold winter and we were much higher than in Ibthorpe – we could even get perfect TV reception by using a coat hanger instead of an aerial – but we were cut off by snow-drifts at one point. We had snow at lambing time but not much and we were only lambing about three hundred ewes. We had no buildings however and only the horse box for shelter – and it rained. A little boy who came to see the lambing lost his wellies in the mud and stated very firmly, "I won't be a farmer when I grow up!" Often on Friday evenings we would have to collect Andrew from Newbury station and Peter at Andover.

We didn't have enough money to do all the alterations immediately so decided to 'make' Peter's room upstairs from the landing by moving the wall in our bedroom and erecting a wall in place of a curtain that was there. We bought an old caravan for Andrew, which we parked at the back of the house. We were still having trouble with the planning department but after six months we were able to move in. There was still work to be done but basically the wall between our bedroom and the landing was moved, which gave Peter his bedroom, and we had the sitting room dry lined as the house had no cavity walls just a single layer of brick. We managed with the original minute kitchen – Dave's Dad built us a wooden lean to for the freezer and washing machine – we stored furniture with various friends and began our six months camping out.

At weekends our family of four was often supplemented with either Rosemarie or Brenda. They were both farmers' daughters and fitted in well and had expertise in handling bottles at lambing time. Peter had met Brenda at University but her home was Kendal in the Lake District. She was reading Nutrition at Guildford but unfortunately when she finished she couldn't find a job so ended up working at Debenhams in Guildford where she worked until after they were married and expecting Bethany, their first child., whichever girl was there had Peter's bedroom and a camp bed was erected in the sitting room for the respective boy. Until the extension was finished we were a bit crowded. We also had the piano to find a place for so it had to live in the kitchen when it was completed. Peter still has it and Bethany, his daughter, has learnt to play. Rosemarie was living and working as a vet's assistant in Marlborough but her home was at Ashford near Barnstaple in Devon. Dave's Dad was a frequent visitor and was always looking for alterations to make or building something, he was a great gardener too, which unfortunately, Dave isn't.

32.　New Year Camps

BARN CAMPS HAD COME TO AN END but we now took the young-sters away over the New Year. About eight miles away is an old style junior school at Oxenwood that has been adapted by the Education Authority for residential use for about thirty people. There was a large dormitory for the boys while the girls slept in bedrooms in what was the caretaker's house, sleeping about twelve. Helpers slept in little odd rooms here and there, there was a well-equipped kitchen and a big hall for eating in and activities. One year it snowed, which added to the fun.

We usually went to Swindon for the Saturday after breakfast and after tea played games, sang silly songs and different people performed sketches, all great fun. If we were actually there for the New Year we would stay up to see it in. We would hold a short service, thanking the Lord for the past year's blessings, concluding with the song "Bind us together Lord" before wishing each other a happy New Year. A lot of practical joking went on, like the time when the girls hid alarm clocks in the boys' dormitory set to go at different times during the night!

Alf and Dora lived next door. His sister and mother had lived in 'our' house but they had both died; we knew Alf and Dora and they were thrilled that we had moved in next to them. It made quite a change in their lives as we put down roots and had all the various guests to stay. They had used both gardens before we moved in; Alf loved gardening and usually won quite a few prizes at the Flower Show. They were both members of the little

Methodist chapel just along the road and Alf loved singing possessing fine tenor voice.

We were so pleased when the house alterations were eventually finished and we could spread out a bit. The extension gave us a large kitchen/ dining room and utility room where we keep the freezer and washing machine. Andrew was able to move into the original dining room, which gave him a small but useful bedroom, and Peter had a small bedroom upstairs. It was still a very small house but has been so warm and comfortable. We bought a second hand Rayburn, which heats the water plus the kitchen, and runs five radiators; Dave and the boys dug the footings for the extension and other jobs they could do to help. Peter wired up the electricity and we all became busy with the paint brush. Although we only had a back yard behind the house we had quite a large front garden where the rhubarb bed became prolific. Alf and Dora's rhubarb was just as good; Alf had a little rhyme he'd recite whenever rhubarb was mentioned:- "Rhubarb tender, rhubarb tough, thank the Lord I've had enough!"

I always enter my rhubarb for the Annual Flower Show, which is held in July. This is always a focal point for the village and a lot of 'looking over the hedge' takes place. I love the atmosphere on the Saturday morning on which it is held when entries have to be taken to the Recreation Ground to the marquee, which has been erected. We unload our cars and carry the entries the hundred yards from the car park trying to be careful. Then we have to find the place where our different entries have to be staged. I lay the four stalks of rhubarb carefully and place the right entry card with my number on it. Then I hunt among the produce entries for the fruit flan; so far there's only two other entries both very different from mine. The scones are arranged around one of my best tea plates and look too big next to the others already there. That just leaves my rhubarb tart (pastry top and bottom) and there's only one other entry so far and only half an hour before entries close. I surreptitiously sneak a proper look at the opposition and then look at some of the beautiful flower arrangements and marvel at the ingenuity of the interpretation of the subjects given. It's nearly ten o'clock so I must leave now.

My daughters in law and grandchildren often used to come over for the afternoon. There's always a lot going on; a dog show is always popular as is ferret racing. We probably won't enter our dogs as they won some rosettes at a village fete recently and didn't think much of the afternoon. There are usually pony rides and one year there was carriage rides pulled by Newfoundland dogs. The school children perform some country dancing before the children's races begin. The Hannington Brass Band performs throughout the afternoon and the sun nearly always shines providing a special atmosphere. Eventually I make my way to the marquee to see if I've won anything. There's a blue card on the rhubarb – a second, not bad. I make my way to the other side of the tent where the produce is displayed, a red card adorns my fruit flan – a first. Along the table and I have a second for my rhubarb pie and finally another first for the scones, which look enormous beside the other dainty entries. I gather up the cards and go to the treasurer's caravan to collect my winnings, which amount to two pounds fifty! One year I won a cup for my fruit flan, another year I gained a cup for a shirt that I had made for David.

33. Alf and Dora

ALF WAS ALWAYS FULL OF FUN and loved to tease Dora. When Alf was 'helping' us move in, he unwrapped some pictures.

"Eileen, can I borrow this picture, I want to show it to Dora?"

"Yes, of course you can."

I trailed him as he went through the garden gate.....

"Dora! Look at what Eileen's given me." He held out the picture.

"Oh Alf, that's nice; I've always liked dogs. Where shall we hang it?"

He owned a Lambretta scooter and the pair of them would spend summer days travelling around the countryside on it, they even went as far as Bournemouth. They often went to Reading where Alf's sister Dorothy lived. Alf was supposed to take her some bean plants on one occasion but when he arrived there found that he had forgotten them; he turned round and came all the way home to fetch them, then returned to Reading. They were lovely neighbours to have; if Dora had cooked a 'figgety' pudding (spotted dog), Alf would bring a slice in for David! I was able to take them into Andover to do their shopping when the village grocers shop, which Roly, their nephew ran, closed down so they were glad that they could come with me

They had seen many changes in the locality having lived all their lives within five miles of Hurstbourne Tarrant. Alf was born in a cottage just down the road and Dora came from Vernham Dean, about four miles away. Alf had worked for the

Forestry Commission and loved to point out large trees that he had planted as saplings changing the landscape from what he had known as a boy. He was exempt from serving in the forces during the war because of his work. Sometimes he had to go away from home and while Christine, their daughter was small, Dora was able to go with him. They would stay in a hut on the site, rather like the hop pickers, Dora said that it was like having a holiday.

Before she was married she was in service at one of the big houses locally and recalls a small boy coming to the door with a message,

"Please Missus, your house is on fire!"

Her 'lady' was playing cricket at the playing fields so Dora sent the boy off to find her while she phoned for the fire brigade and gathered up the dogs and went outside. There were flames and lots of smoke coming from the chimney and it was a long time before the clanging bell of the fire engine could be heard by which time the owner of the house had arrived. The fire was quickly dealt with and Dora had the job of clearing up the resultant mess.

"Eileen, could you make an appointment with the doctor for Dora?"

"Alf I will, but what's the problem?" They were in Scotland on holiday with a coach tour so I thought it must be urgent for Alf to 'phone.

"The doctor here thinks that she's had a heart attack and she's not at all well."

"O.K. Alf, I'll do it straight away, give my love to Dora."

I managed to get an appointment for Tuesday and took them for it. It was a heart problem; she was put on medication for it and told to take things easy. She soon bounced back to her usual placid self but found that she couldn't do what she had been used to doing. Alf took over the washing but couldn't do the ironing; He managed the twin tub washing machine so well that when it came to the end of its life he bought an identical one; he wasn't too good at hanging the washing out though so I'd often do it for him. He extended his cooking skills and could manage the basic things; he could even cope with a fruit cake. He was diagnosed with diabetes but I found a fruit cake recipe especially for diabetics so he was able to make and eat that.

He found that although he had to do Dora's work as well as his own he could manage quite well. At blackberrying time he would scour the hedgerows for blackberries and as he had a big apple tree in the garden learnt how to make jam. He also made sloe jam – with what he called 'bullisons'.

They loved to go on bus journeys so one day set off for the bus stop to go to Newbury; having reached there they caught another bus to Basingstoke. They were to have caught another one to Andover and then finished the journey with one back to Hurstbourne Tarrant.

The telephone rang. It was Alf.

"Eileen can you come and get us, we're at Basingstoke Hospital. Don't worry Dora's been checked over and she's all right but she fainted." They were both sitting on a seat outside the hospital, Dora looked quite white but once she had arrived home and had a cup of tea she felt fine. They were well into their eighties by now. Alf developed heart trouble and went into hospital where he died but Dora carried on to her nineties with the aid of part time carers who came twice each day. She also had meals on wheels and enjoyed the food that she hadn't had to cook. I was her 'carer' when there was no one on duty at night.

"Dave, quick someone's ringing the doorbell." I looked at the clock; it was 2 a.m. Dave put his head out of the window, there were two policemen outside.

"Is Mrs Sullivan there?" I was halfway down the stairs trying to put on my dressing gown.

"The Carer's service has rung us up because Mrs Knight activated her 'Lifeline' but didn't say anything. Your name is the one we have for out of hours care; is that right?"

"Yes, officer we had better see what's wrong."

Followed by the two policemen, I went through the adjoining gate. I opened the door.

"Would you like us to go in first?" asked one of the policemen.

"No, because if she woke up and saw a policeman it would frighten her." Her bedroom was on the ground floor; we crept in. she looked asleep,

"Dora, are you all right?" She opened her eyes.

"What are you doing here Eileen?"

"Dora, you pushed your lifeline by mistake and these policemen were sent to make sure you were all right."

"That's nice of them. I'll go back to sleep now."

"Madam, are you sure that nothing is wrong?" One of the policemen asked her. Only after she had verified the truth of that did they turn to go.

Dora didn't believe me when I told her later that she had had two policemen in her bedroom in the middle of the night!

On Sundays she enjoyed going to the little Methodist Chapel in the village; her nephew Roly came to meet her from chapel and took her back for the rest of the day. We collected her from Roly's and took her to the Gospel Hall with us on Sunday evenings, which she enjoyed as her sister, May, Roly's mother, who lived near the Hall came too. I acted as Dora's carer on Sunday evenings as by the time we reached home it would be too late for one to come out. She went into the bathroom one evening to put her teeth in their container.

"Dora, where's your hearing aids?"

She felt in her ears.

"I don't know."

I found them in the bathroom in the container with her teeth and managed to dry them out so that they worked again.

She had to have a hernia operation when she was nearly ninety but bounced back again although she was in hospital for a week, which she didn't like and was glad to be home again. She was an amazing person and everybody loved her. She became increasingly frail and spent a few years in a retirement home in Andover but then was transferred to a nursing home near her daughter at Witney.

The retirement home in Andover where she stayed was the same one that Mum had been in. A lady, Kitty, from one of the local churches had the idea of having services once a month for any residents that liked to come. We would go to the various departments, Ash, Oak, Willow and Beech, and ask the ladies if they would like to come. There'd be quite a few of them and perhaps four of us helpers. One lady played the electric organ and we would sing a couple of hymns, have a Bible reading and Kitty would speak to all of us of God's love. It helped to fill the residents day up and in some cases was a real link to their past.

34. Ted and Lily

AN ELDERLY COUPLE MOVED INTO THE VILLAGE, Ted and Lily Frampton, and joined us at the Gospel Hall. They had been Army Scripture Readers, which is an organization set up to reach out to the soldiers with the Gospel. Help is given in both practical and spiritual matters and the soldiers are welcome in the Scripture Readers homes. Ted and Lily were now retired and their first problem was to find somewhere to live as their house went with the job. They prayed about it and decided to take the first house that was offered. On the same day that they heard from the council in the London area in which they were living, they also had a letter from a Trust that were inviting them to see a cottage in a village that they had never heard of – Hurstbourne Tarrant! They would never have thought of living in a village, as their original home was Belfast. Ted had worked in Harland and Wolff shipyard until he was marvellously converted to Jesus Christ in his thirties. He had worked over in Belfast at Sands Soldiers Home until he felt that God wanted him to work for S.A.S.R.A.

It was similar to being an Army chaplain but he wasn't employed by the Army but by S.A.S.R.A. This organization, Soldiers and Airmen's Scripture Readers Association, is funded by donation; Ted and Lily had worked for them for some years and had travelled fairly widely, with postings to Aldershot, Catterick, Aden and Singapore where their sixteen year old only son died.

The accommodation offered them in London was a seventh floor flat and in a rather deprived area. Friends had to bring them down to see the cottage as they had no transport of their own. It was a cold December day and the journey seemed to take for ever, they had arranged to meet the trustees at the cottage. They eventually found it, a real 'chocolate box' thatched cottage with a stream running along the front of it. Ted was quite tall and kept bumping his head on the low oak beams across the ceiling. It was very small with the front door opening straight into the sitting room, which had four doors opening from it. One led to the galley type kitchen, which in turn had a door leading to the bathroom. Another door from the sitting room led to the box stairs, which were short and steep with a turn towards the top. The fourth door led outside to the large back garden. Upstairs there was one large bedroom and leading from that a small room, which was above the kitchen.

The views from the windows were superb with a view of the garden with hills in the background vying with the view from the front window of a farmhouse, farmland and the parish Church. The garden was long, mainly lawn but the 'shed' seen from the window turned out to be a summerhouse where a previous owner had had a little nursery school. The blackboard was still attached to the wall and some of the little chairs remained. Ted and Lily used it, after refurbishment, as an extra bedroom as they were very hospitable and most weekends they would have guests. Some of them would be service men and many a romance began over the washing up. Ted and Lily had never lived in the country or even visited anyone who lived there, so living in Hurstbourne Tarrant was a new experience for them.

Ted had no gardening experience but was keen to try his hand at it. When the weather became suitable he dug up the garden at the end of the property for vegetables, and attacked the flower beds with enthusiasm. He loved to show off his garden and when the vegetables flourished he'd send his friends back to London loaded with them; new laid eggs were available from the farm over the road. They had no car when they first came so were dependent on us for transport but as they had plenty of visitors they didn't feel cut off. A stream ran along two sides of the cottage and one year it was very high and we had to put sandbags in front of the door in case it flooded.

When it was nearing the date of their Golden Wedding Anniversary, some of us decided to give them a party, it snowballed and eventually a hundred and thirty guests joined in a service of thanksgiving in the Gospel Hall followed by a reception in the village hall. We had managed, without realising it, to invite someone from every posting they had been to. There was about half a dozen friends from Northern Ireland and Lily's sisters from Canada and Australia. We had to tell Ted and Lily why they had come as obviously her sisters wanted to stay with them but we kept from them the scale of the venture.

The Gospel Hall was packed with some people having to stand at the back. The singing of 'Great is Thy faithfulness' nearly lifted the roof off; we had asked another friend to sing a solo and various others did the prayers. The Village Hall was only a short walk and as the weather was nice, people were able to sit outside to eat. Peter and Brenda were mainly responsible for the food, aided by the ladies from the Gospel Hall. It was a sit down affair, we had hired crockery and cutlery, the village hall was decorated with golden teddy bears and altogether it was a momentous occasion. Speeches were made video recordings of friends in Singapore were played and Dave brought everything to a conclusion with a song we had written about their life with everyone joining in the chorus of 'We love you Ted and Lily, glad to share your great day'.

35. A Trip to Normandy

WHEN WE RETURNED FROM OUR HOLIDAY from France that year we told Ted and Lily about the war cemeteries we had visited along the Normandy coast. Ted was very thoughtful and said, "I wonder if it would be possible to find my mate's grave. We were not allowed to bury anyone who died just had to make sure they still had their 'dog tag'." Ted was in the first wave of soldiers sent over on D-day and when his mate died beside him he could only follow instructions. We decided to take them both over to France if we could.

When we told friends at church they started to give us money towards the trip, which was to be for four days in September. We rented a gîte and after an enjoyable ferry crossing arrived quite late in the evening. We soon had the kettle on, I had taken a teapot with us as gîtes aren't equipped with one and after a drink, we were soon in bed. We were up early the next morning and decided to go to the big army cemetery at Bayeux as Ted wasn't sure where he had actually come ashore. It was a hopeless task there as the 'walls' of names were so vast that it was impossible to read them.

We went on to the very interesting museum at Arromanches where an artificial harbour was made to get vehicles ashore. Concrete blocks were towed across from England and sunk to form breakwaters and some of them are still visible through the windows of the Museum. We've discovered that the French have a real flair for displays of anything and the museum was excellent. We had lunch sitting high on the cliffs beside a big gun emplacement in front of a field of sweetcorn, which fascinated Lily as she had never seen it growing. We tried another cemetery and learnt there that in the gatehouse of every cemetery is a book containing the names of everyone buried there. We sat Ted down with the atlas to try and jog his memory. "Coalvill! That's the place but I didn't think it was spelled like that, Colleville." There were two places bearing the same name one had the suffix 'Montgomery' and the other 'Bas'. Ted was sure it was Montgomery so we decided to go there the next day.

It was a grey damp morning. We eventually found the place, which was a quite small cemetery compared with the others we would see. In the gatehouse was the treasured directory. There were a number of Irish names in one place. "That's him" Ted said stabbing his finger at the page, "we used to call him 'Ginger' for obvious reasons."

We wrote down the number of the grave. Soon we were standing there in front of it. Ted was overcome with emotion. Dave and I moved away so that Ted and Lily could share the poignant moment together, we were glad we had come.

We were going home the next day but had decided to visit the Caen Memorial Museum on the way to catch the ferry. David and I had been before and it's a very awesome place.

The route inside leads down a big spiral going downwards with films of atrocities appearing and Hitler's voice booming out from the depths. There were films shown of the actual D-Day landings at various places and how everything came together at the right time. There were models of the ships bringing the troops and the appalling losses they suffered. Exhibits about the holocaust were very moving and in one place there were hundreds of lights representing lives that were lost.

We had our lunch there and had a rush to catch the boat home again. It had been a brief but worthwhile four days and much appreciated by Ted and Lily.

36.   Goodbye to Old Friends

LILY SUFFERED A NEARLY FATAL HEART ATTACK a few years later, and had to be resuscitated to get her heart going again. It was brought home how near to death she was when Ted gave me her clothes to dispose of, which were all cut from top to bottom down the front. Her clothes had had to be cut off her to allow the resuscitation team access to her heart. She was in hospital for some time and had to be very careful when she came home.

Ted died in 1996 while they were on a visit to Ireland. Lily wanted him to be buried there as she was sure that was what he would have wanted. He was able to be buried in the same grave as his sister in an enormous cemetery on top of a windy hill. Dave and I went over for the funeral and I was going to stay on for a few days and bring Lily home. We were picking up Tim, another friend and had to run to catch the plane! The funeral was held at Shankhill Road Baptist Church and I found it quite scary to see the armed policeman outside the Church as there were still problems in Northern Ireland. Lily and I stayed with friends in Belfast and from the upstairs window we could see the Harland and Wolfe cranes in the dockyard. David stayed one night in Belfast then had to go home to see to the sheep. Lily and I were there for another three days before catching the plane home. Ian Paisley was on the same plane as we were.

Lily continued to live in the cottage as she loved living there. Friends still came to stay with her and she had a trip to her sister in Australia and also went to visit friends in Singapore. I used to spend one morning each week helping her with the garden but it became too much for us so she had to have a gardener for an hour every week. She was able to visit Elizabeth, a particular friend who lived in London with her family; I'd take her to catch a train in Andover and her friend met her at Clapham Junction. I was going to meet her after one such visit and asked Dora if she'd like to come to Andover and have lunch in Tesco's with Lily and I on the way back as we sometimes did. We met Lily and drove to the shop. We decided to use the toilet before lunch so were pleased to find it empty. Dora and Lily occupied the two cubicles, Lily was soon out but Dora had a problem.

"Eileen, I can't get out" came Dora's voice.

"What's the matter Dora?" I replied.

"I can't find the bolt!"

We realized that something was wrong. Lily climbed on the seat of the loo in the other cubicle to try and see over the top of the wall while I got down on the floor and peered under the door. Just then, the outer door opened so Lily climbed down from her perch. The visitor stepped over me, used the toilet, stepped over me again, and washed her hands without saying a word! I realized that Dora was facing the wall of the cubicle rather than the door so I reached under the door and tried to reposition her feet. I managed to get her to turn them but she just turned straight round again! At this point Lily disappeared returning minutes later with a porter. He climbed onto the washbasins and peered over the top of the cubicle,

"Look out dear, I'm coming over!" he called to her. She couldn't make out where the voice was coming from. He lowered himself down and opened the door much to our relief. Dora was completely unconcerned about the whole event.

"Can we have our dinner now Eileen, I'm starving."

We were soon tucking into our meal and a welcome cup of tea.

Lily coped quite well by herself but had to have blood transfusions every few months and became increasingly frail. She had cataracts in both eyes removed and then developed a hernia, which necessitated an operation. Eventually she was finding it difficult to cope with living in the cottage so went into a lovely Christian Residential Home where she lived for two years before a final illness in 2008. She wanted to be buried with Ted in Belfast and we had the job of arranging that and also a Memorial Service, which was attended by a hundred and fifty people. She was greatly missed.

37. Special Needs Class

IN 1985 I FOUND A VOLUNTARY JOB in Andover helping with a class of about ten Special Needs Adults. I had no experience but found that we related to each other quite well. After a couple of years helping I was able to go on a course to get a qualification and took over the class when the tutor left. Another class was amalgamated in with mine and there was approximately one voluntary tutor to two students, there were about twelve students. The students were all ages; the eldest was in her fifties and the youngest in her twenties. Some of them had lived in old type institutions where there was little stimulation. Each student had their own learning package and we often went out into the town to study various things. We spent one morning at the Police Station where the sergeant locked a couple of them in a cell. They had their fingerprints taken plus their height checked and really enjoyed their trip out. At Christmas we managed to take them for a meal at one of the pubs that catered for parties, they were always well behaved if a bit noisy.

One year they came out at lambing time, which was certainly an experience we'll never forget. One of the helpers had brought a video camera with her to record every moment. David had to deliver a lamb as it was coming backwards and he had to turn it round in the womb. Time was running out and as we were due at a local pub for coffee, we had to go. As we were walking away, the lamb was delivered. Dave held it up and the event was duly captured on camera. What the students didn't know was that the lamb was dead as Dave had made its ears waggle!

We took them on a visit to a butterfly farm and also a Zoo up in the Cotswolds when it rained the whole day. There was a good restaurant and they spent a lot of time there eating chips.

Eventually the class was closed down as the policy with special needs students altered but it was a happy five years. I often see the students in Andover and they usually remember me. One chap used to come out on the bus once a fortnight to 'help' Dave. It was amazing as he would look at the milometer in the Land Rover and tell Dave how many miles he had done since he last saw it. He was very fond of nature and was quite knowledgeable about it. His ability to read and write was minimal and his attention span limited. He watched all the wildlife programs on the television and would quote numbers when he came to class such as the starling population of New York!

Most of the students lived in flats with two or three sharing although there was one group of eight living in a house with a live in carer to be on hand when needed.

38. Sheep Farmer Again

GRADUALLY DAVE'S SHARE FARMING came to an end and he had to find something to replace it. He decided to buy sheep as 'yearlings' – sheep that were a year old and sell them on after breeding as in-lamb ewes. This worked for a few years and then the bottom fell out of the market and he was left with three hundred ewes unsold. There was only one thing to do, he'd have to lamb them himself and have a ewe flock.

This meant rethinking the whole situation. He rented a barn that he could use and he'd be able to buy hay and straw off the field. He bought half a dozen rams from Wilton sheep fair and he was in business as a sheep farmer again. It worked quite well and Dave used his barn, getting them in before lambing and feeding concentrates and hay. He realized soon however, that it was a costly way of doing things. The lambing percentage wasn't too bad but Dave felt that he could do better. There seemed to be no shortage of grazing available at a reasonable rent so the next step was to arrange his lambing around that.

He reasoned out that if he lambed in May, the grass would be growing so they wouldn't need so much bought in feed. The days would be longer and warmer, which would mean that they would be able to cope better and also most farms would have finished lambing so he'd be able to get help.

Andrew and Rosemarie were married and managed to rent a cottage in the village. Andrew and David set up a fencing business between them, which continued for about ten years. They specialized in electric power fencing as there was a demand for it especially with a nearby farm that kept pigs and needed fences moved every few weeks. Philip was born and we all loved him. I usually looked after him on Saturday afternoons so that Rosie had the afternoon to get things done. Andrew and Rosemarie eventually moved to Tidworth as they were able to buy a house that had been de-commissioned by the M.O.D. where Andrea was born. Philip is now a strapping six-foot-four giant working at Lloyds TSB and Andrea has the job she wanted working at a boarding kennels.

Peter met Brenda, also a farmer's daughter from Cumbria, at Surrey University and married her three months after their graduation from University. They settled in Aldershot, as Peter had found a job with a commercial firm at Farnborough but eventually obtained a post with Surrey County Council at Kingston on Thames. Bethany, their daughter, is eighteen now and studying at Sixth Form College. Daniel is fifteen and helps David with the lambing whenever he can. Michael is thirteen and loves cooking. The Hairy Bikers are his heroes.

39. Lambing

DAVID HAD TO DEVISE A NEW SYSTEM for lambing as he was limited in his options for locality. One farmer from whom he rented land had three adjoining fields which seemed ideal for the purpose for which Dave needed them. He worked out that if he used two fields during the day he could move them into one of the other fields at night with pens in one of the day fields. The pens were to bring them into after they had lambed where the ewes udders would be checked, the lambs were castrated and tailed with rubber rings applied with a gadget called an 'elastrator'. The navel was treated with iodine to prevent infection and the lambs checked to make sure they could suck adequately.

As Dave had a thousand ewes to lamb he employed somebody to help, often one of his previous students would be available and would move in with us for the duration of lambing. Because there could be two or three people helping at different times, Dave devised a system of keeping track of at what stage the sheep were. He found the lids of certain containers were excellent for the job and he marked them with four numbers each relating to a different job. 'One' was one navel, 'two' was two teats, and 'three' was three square meals 'four' was to turn them out. It was a system that worked well; the discs were attached to the front of each pen and moved round until all tasks had been done. The ewe and lambs would be removed from the pen after David was satisfied that both Mum and babies were fit and well. They were then moved to 'nursery paddocks', where there was a clean area of fenced grass where all the 'families' from one day were put and could be monitored for twenty four hours to make sure that they had bonded and after that they were taken by trailer away to a field. The pens were cleaned out and fresh straw put down after each occupant so cross infection was kept to a minimum. David always left the ewe with her lambs on the birth spot for some hours, just checking that the lambs were suckling satisfactorily and the ewe 'cleansed' properly. He found that the ewes liked to find a quiet spot with some shelter so he left clumps of 'designer' nettles around the edges of the field, behind which they could give birth. He would leave them there until the other ewes had been fed their concentrates then he would walk them gently down to the lambing pens.

The ewes had numbered ear tags and if she had only one lamb, Dave wouldn't number the lamb as she was unlikely to lose it. If a ewe had twins then Dave would number the lambs with the same number; he had branding irons, which he used with special paint that wouldn't wash off straight away. If she had triplets he would paint a 'T' on one side with its number on the other. If the weather was particularly bad, he would bring any ewes that needed attention into a pen before they gave birth. A ewe is very restless before the lambs are born and will pace up and down looking round towards her rear end, she may lay down more than once and begin to strain. After a while her pushing becomes more urgent and gradually the head of the lamb appears followed by its body. Sometimes it's still occupying its birth sac, which the ewe will lick away before starting to wash the lamb vigorously all over. This not only cleans the lamb but also stimulates the lambs breathing and warms it up. Within a few minutes the lamb totters to its feet and struggles to find its milk source. Soon it latches on and its tail wriggles with satisfaction. More often than not there is a second lamb and the ewe has to leave the first one to see to the new arrival.

Dave is always checking round and would notice if she needed help. This is sometimes given if the lamb is coming out in the 'breech' position when he pushes it back inside the womb and turns it round. David has huge hands, which are very strong and can manage this sort of job easily. He leaves them on the birth spot if the weather is fine until he clears the field at the end of the day when they are transferred to an individual pen. Sometimes a ewe has triplets and he tries to foster one of them on as it's difficult for a ewe to manage three lambs. If he discovers this at the right time he can put the adoptive lamb with the live one, which is still struggling by the ewe and smother the other one in the amniotic fluid, which she then proceeds to lick off. We are sure the sheep is puzzled about how the second lamb arrived there without her having to push!

If a ewe has a dead lamb and another ewe has triplets, David will skin the dead lamb and put the skin on one of the live lambs, which the first ewe will usually adopt. After a while the skin gradually stretches and trails behind the lamb like a dressing gown. Any lambs that aren't fostered end up at home having to be bottle fed every four hours although after a few weeks we wean them on to a 'lamb-bar', which is like a bucket with six teats attached near the top with tubes leading into the milk. These teats have non return valves so the lambs don't choke themselves. They don't like it at first and it's a time consuming job persuading them how much better it is for them.

We use cold milk as that can be left in all day and the lambs have it as and when they want it. Eventually they give it up altogether and can become one of the flock.

We usually had whichever person, usually a girl, was Dave's student to live in, as it was so much easier when we were very busy and Dave needed to work a shift system. Louise was one of the most memorable. It was a year that was kind to us and we had good weather in April. Louise quickly decided that shorts were the required dress for lambing and although sometimes she'd get filthy, a bath soon put that right. Towards the end of lambing when the arrival of the lambs tailed off a bit she decided to paint David's sheep trailer so she'd be on top of a ladder with green paint everywhere. The girls had much more of a 'feel' for the sheep than the boys did – and more stamina too. Once the ewes had lambed they were ferried to either a field of twins, one of singles or one of triplets. We've only ever had one set of quads and they didn't all survive.

It is so easy lambing in May. The weather is usually a lot warmer and the days longer, which makes shift work more pleasant. When lambing is over there is always a big clearing up session as most of the bales have to be burnt and all the equipment put away. There is usually a couple of sheep left to lamb so they would be moved to a little paddock behind the house. Also any odd lambs that can't be mothered join them so I can cope with bottle feeding them. There is inevitably one at least that has to live indoors for a couple of days – that's long enough; lambs can't be house trained. When they eventually join the flock, for some time they will run to anyone who enters the field and they never lose their trust of humans. It's always sad though when a lot of time and energy has been put into rearing a lamb if it subsequently dies.

40.   Cuckoo Lambs

BRIDGET O'SULLIVAN WAS ONE OF THOSE. Dave discovered after her rather traumatic birth that she only had one eye. Realising that she wouldn't be able to compete for her mother's attention he decided to bottle rear her. Any lambs that are bottle fed end up with a name, which they quickly learn. His reasoning for giving her such a mouthful of a name went as follows: – Sullivan means 'one eye', 'o' is of, and the name 'Bridget' seemed to suit her. She was called Biddy for short! These lambs that are hand reared are known as 'poddy' lambs 'cade lambs', 'cuckoo lambs' and 'pet lambs'. They are difficult to rear as they have to be taught to first suck from a bottle, which is strange to them; we have to pay attention to hygiene as they pick up diseases easily. Then as they get older they want to walk right behind you and generally are a nuisance.

As well as Bridget I had Widget, Midget and Twiglet to feed. They were three who for various reasons had been rejected by their mothers. Bridget didn't think much of them and kept thinking of other mischief to get into. She raced across the paddock one morning but seemed to be dragging something behind her. It was a nightie of mine that had been on the washing line and somehow she had her head through a strap! After all, people wear clothes so why shouldn't she? She loved 'helping' me and rooted around in the clothes in the washing basket so some things got wetter than when I brought them out. She tried her hoof at digging the foundations for the new garage that David was building; Bridget, Widget, Midget and Twiglet spent happy hours with Dave as he was digging. David now keeps an old tractor in the garage, which he's hoping to renovate but he never seems to have the time.

We have other unusual animals and birds around. Bertha the bantam came to live with us and although she shared us with a neighbour she liked living in the garage. She spent a lot of time on David's work bench; we think she was trying to hatch a screwdriver. She didn't succeed however; Dave could have used some small screwdrivers. She learnt that if I went into the garage I'd remove her so fluffed herself up to twice her size and began to scold me. Once I'd picked her up and put her outside with some corn she'd decide that I was all right but still kept up a constant muttering. I couldn't find her one morning; when I finally located her she was in the bag of chicken food sitting on an egg! It was the only one that she ever laid.

The first pet lambs I had when we were in Kent I found difficult as it was a new experience for me. I had four of them including Larry. One morning when I went to feed them I discovered that they were all black; Dave had swept the chimney and put the soot in a heap in the garden! Larry was very tame and would follow me down to the shop. I think Miriam was our favourite sheep. Her mother had rejected her, ewes did that some times and she was hand reared. Her favourite occupation was going to work with Dave and walking right behind him when he walked along the road with a couple of hundred sheep behind her. This type of sheep was known as a 'Judas' sheep; they would follow her into trouble everywhere. She'd walk in the middle of the road – on the white line if there was one. She loved coming indoors and never disgraced herself, it is impossible to house train them but she was always very good. She discovered where the sack of dog biscuits was and could be seen with her head buried in its depths.

'Pet' lambs have always been a problem. Some of them are fostered on to ewes whose lambs have died. Ewes reject a lamb if it's just put with her so Dave has learnt to employ strategies to overcome this. If Dave has had to help a ewe and the lamb is dead he will hastily fetch an orphan, smother it all over with the birth fluids, tie its legs together so it wriggles and present it to the ewe. Often it works as by the time she has licked it off she really thinks it was hers, although she's a bit mystified as she's sure she only had one lamb and Dave can untie it.

Other times he can employ a similar idea if the lamb dies after birth but the ultimate is when he will skin the dead lamb and tie its skin on the live lamb. He removes the skin carefully leaving 'cuffs' at the leg openings. He then fits the coat on the live lamb and the mother can smell the dead one and accept it. After a few days David removes the coat at arm's length, as it's very smelly by now, and the ewe is quite happy with her adopted lamb.

Failing all else Dave will put the ewe 'in the stocks' – two stakes banged into the ground, the ewes head fits between them and the stakes are tied at the top to stop her escaping. She is then given a lamb to feed. She can't get her head round to reject or bunt it and after a couple of days Dave will release her and she'll stay in a pen for a while to make sure that the lamb is feeding satisfactorily. She's had access to food and water as her head is free enough to move up and down and she can lie down comfortably.

David's aim is to have no pet lambs left at the end of lambing but in the weeks following the end there's usually some that are going to have to be hand reared. They are scrawny little things and need constant monitoring and feeding. They tend to get pot-bellied as until we can get them to use the 'lamb-bar', they have to have five meals per day, but feeding them by hand is no substitute for ad-lib feeding. They have to have names so we can distinguish between them. They often get named after film characters or nursery rhyme figures. Katie was one such lamb and proved to be a real character. She was born about the time that 'Titanic' was released so was named after Kate Winslett. A ram lamb was born and abandoned by his mum soon after so became Leo!

My role in lambing is minimal, but the odd lamb ends up in the bottom oven of my Rayburn until it starts staggering around the kitchen. It's miraculous how quickly the gentle heat revives them and after a bottle of milk they are ready to go back to their mothers. If any are orphans they get corralled next to the Rayburn as there's a little cubby hole there where the coal hod usually stands. I line it with plenty of newspaper and find a cardboard box and an old jersey and it makes a cosy bed. Once they can escape then they're out of the house, David does find mothers for most of them.

David keeps the lambs in his post lambing fields until he's satisfied that the ewe and lamb(s) have bonded well; then he loads them into his sheep transporter, which has sections for individual ewes and lambs and takes them to another 'clean' field some little distance away. Having no land of our own Dave has to ask other local farmers if they have any spare fields. All the local farmers get on very well together and help each other out whenever they can. Dave will help them with some of their sheep jobs as most farmers can no longer afford to employ a full time shepherd.

41.   Trailer Rides

DAVID HAS FOUND ANOTHER USE for his big sheep trailer. We help to run a children's Gospel Club at the Gospel Hall and the opening club night of the autumn term is a tractor and trailer ride. We leave at six o'clock, with bales of straw along the middle of the floor of the sheep trailer, about two dozen children plus about ten parents can fit in easily. Andrew hitches on a powerful tractor from his work and with Dave riding 'shot gun' we set off. It is the most uncomfortable and noisy means of transport that has been invented; the children love it but don't stay seated for long and jump up and down, squealing all the time.

We soon leave the road and take to the cart tracks eventually arriving at the beauty spot of Coombe Gibbet, high up on the Wessex downs. The view from here is stupendous, looking one way towards Oxford and the other towards Andover and beyond. The children run around and as there is a replica of the original gibbet there they run up and down the mound on which it stands.

After a while we get the food out, thinking that we've brought too much but the crisps, sandwiches, sausage rolls and cake soon disappear. Then it's back to running around for a bit longer and a search for suitable bushes, girls one way boys the other. We climb back on the trailer again and have a quieter ride back as the children are so tired.

We arrive back in Vernham and the parents, seeing us pass, leave their drinks on the outside tables of the George where they have been waiting for the children. One year it had rained all day and I felt sure that it was too wet but Dave never gives up.

"The trailer's got a corrugated iron roof, it'll be fine!"

I ended up wetter than I've ever been in my life. Yes, there was a roof but it consisted of overlapping sheets of corrugated iron. When we went up a hill one way the rain flowed through the gaps; going the other way it came at you from a different angle. We all looked like drowned rats at the end but the children thought it was great fun and stood at the front of the trailer where there was the spray from the wheels! We stopped miles away from anywhere to eat our picnic but soon the children were out of the trailer and running around in the rain – it was very cold as well. I was glad to get home and have a hot bath.

42. Philip's Knee

DAVE HAD HIS FENCING BUSINESS as a side-line to the sheep enterprise and when Andrew finished at college he decided to join Dave. The fencing used was the Ridley Rappa system, invented by Harry Ridley a few years earlier. This utilized three wires powered by an electric current passing through from an energizer situated at one end of the wires. The whole system was portable with a special contraption called a 'Ridley Rappa barrow' that unwound the wire as the barrow was pushed along. It also carried the metal stakes that carried the wires. This made it quite easy to move sheep into unfenced fields as the fencing was quickly erected. Some farmers used it to keep their pigs in and Andrew was kept busy as the fencing needed moving quite often. He'd spend a month each year working on a farm helping with their lambing and eventually was offered a full time job there, which he accepted and he's still there now.

He lives at Tidworth with his wife, Rosemarie, twenty one year old son Philip who works for Lloyds TSB and seventeen year old daughter Andrea who is studying Animal Management at Sparsholt and hopes to get a job with dogs. Andrew and Rosemarie are very involved with a small but thriving church at Shipton Bellinger; he has his day off during the week so that he can take R.E. lessons and school Assemblies. He even took a lamb in at lambing time so could speak about "Jesus, the Lamb of God who takes away the sin of the world."

When Philip was twelve he developed a painful knee and had difficulty walking. The doctor referred him to a consultant and various tests were done and cures tried but nothing seemed to be any use and he still limped and had to use crutches. It was even suggested that it might be psychological but on probing questions that was ruled out.

It was the end of the summer term at school and Andrew and Rosemarie had a family holiday fixed on the Isle of Wight but Philip refused to go. I said I'd have him as I have a downstairs bedroom and he was quite happy with the arrangement. I asked if it would be all right if I took Philip to see an acquaintance who practiced osteopathy and they were quite happy with that. My friend said it wasn't her field and referred me to a physiotherapist who was also a C of E minister. We went to her house in some trepidation but she was delightful and Philip settled down on her settee with the cat.

She took a case history first, examining him at various significant points. In the end she decided that the knee was beginning to heal and gave him a series of exercises to do to strengthen the knee. We set aside times each day to do them and at first I felt that they were not doing anything but as the week progressed he was able to do more and more. When it came to the day when the exercise was to walk without his crutches and he did we were both near to tears.

We had made the week a holiday for him and did various things. One afternoon we spent over at a trout farm, fishing. We caught six beautiful trout with a very rudimentary fishing rod and bread for bait; our only problem was that we became too excited and got the line tangled up, which necessitated the proprietor having to sort it out for us. Another day we went with my neighbour and her two children for a trip on a horse drawn narrow boat at Kintbury. I had been on it before and it's a peaceful way to spend a couple of hours. The children were fascinated by the countryside through which we passed and the horse that pulled the boat. We went through a lock, which was interesting and the horse was walked to the other end of the boat to pull it for the return trip. We had taken our lunch with us but drinks were available on board.

The weather was hot and the three of them splashed around in Jack and Jessica's large paddling pool. David had promised Philip a ride on his farm motorbike when he could walk and by the last evening he had dispensed with his crutches so went for a trip round the farm with Dave. It was a thrill for us all when Andrew and Rosemarie came to collect him and he's had no more trouble with the knee since. He's twenty two now and works at Lloyds TSB in Andover and really enjoys his work.

43. Sheep Health

WHEN THE LAMBS ARE BORN, the first job that Dave does is to check their navels and dip the umbilical cord in iodine to prevent any infection. He puts a rubber ring on its tail, this stops the circulation and the tail falls off after a couple of weeks, this is for cleanliness. It's funny to see the lambs feeding from their mothers with the stump that's left circling furiously as a lambs tail wiggles incessantly when it's feeding. He also puts a rubber ring on the testicles of the ram lambs thus castrating them to prevent them from breeding. As we use North Country mules (nothing to do with the equine mules) and a Suffolk ram the lambs are half-bred and to keep the ram lambs for breeding wouldn't be wise.

When they are about twelve weeks old they have their first dose of worming medicine administered by a drench gun that squirts the set dose to the back of their throats. This procedure keeps them free of intestinal worms, which would stunt their growth. The other procedure that's carried out is an injection for various diseases. They are weaned from their mothers at sixteen weeks and have to be out of earshot of their mothers or the ewes would break out of their field to reach their babies. The ewes lose a lot of weight while they are feeding their lambs but have several months before the breeding cycle starts all over again.

David inspects all the ewes and rejects some if they are more than eight years old and past breeding age; they go to the butchers for mutton. He also trims away the wool around their rear as otherwise the wool gets soiled, which encourages flies, which will lay their eggs in the soiled wool, which hatch into maggots. I sometimes wonder why we keep sheep, there seems to be so much that can go wrong with them.

Some diseases that sheep get are transmissible to humans and Dave always has to be aware of this. One of the nasty ones is 'orf', or to give it its full name – contagious pustular dermatitis. It infects the ewe's teats and the lambs mouths where festering sores break out, it is treatable but it's an awkward procedure. In humans it can go on to a 'flu like illness and can take months to clear. We've both contracted it at various times in the sore form, and it's quite hard to shift, antibiotics seem the best option but David is allergic to penicillin as he discovered quite early in his shepherding career. He was given it when he cut his hand badly while his boss and wife were away and he lived on a remote farm on Exmoor. He managed to get to the next farm where he was given a lift into outpatients and had the cut attended to plus an injection of penicillin. Later that day when he was back at the farm by himself, his hand started to swell up, then his arm... His employer returned at that point and quickly took him back to the hospital where he had another injection to counteract the effects of the penicillin.

Of all the diseases that sheep are susceptible to, Foot and Mouth disease is the most feared as animals have to be destroyed if they contract it. When we had the last outbreak in 2001 there was an immediate ban on moving animals within a certain radius of the outbreak. This affected some farmers very badly especially those who had sent their sheep away to winter, The Romney Marsh farmers were affected badly with sheep in the low lying meadows needing moving to the West Country for the winter as the fields in Kent are liable to flood. The deep dykes or ditches that surround the fields fill up and overflow making part or the entire field unusable.

How did it affect us? We usually sell our lambs as 'store' lambs, which means that someone else fattens them and sells them as 'fat' lambs. David had to keep them and fatten them; a neighbour let him use a field of turnips, which saw them through to the right weight.

44.   Farm Crisis Network

DAVID IS A MEMBER OF AN ORGANIZATION called Farm Crisis
Network, a helpline set up by farming people to help others
cope with the problems that arise in the farming community.
More farmers commit suicide than workers in any other type of
employment. Volunteers man the telephone in four hour shifts
answering every call that comes through and referring the caller
on to the best person or organization to help them. David is on
call for his four hour shift on Monday afternoon, some days he
can have four or five calls on others only two or three. A lot of
the calls are about subsidies that should have been paid and
other money problems. These are passed on to the nearest
volunteer who can then visit or phone the enquirer to help them
and follow through any cases to their conclusion.

The Foot and Mouth outbreak made for many distressed calls
where people had had to have their stock destroyed; as it was to
some their only source of income with dairy farmers being the
worst affected. One of our flocks couldn't be moved as they were
within the set quarantine distance of an outbreak. They survived but it was an anxious time.

David also had ewes on the North Wessex Downs, a lovely
area of high downland. The ewes should have been brought
down to fields near home for lambing but they couldn't be
moved. Dave had to shepherd them using his motorbike and
just rubber ring the lambs and leave them out where they were
on a very exposed hill side.

45. Shearing

SHEARING IS AN ANNUAL JOB and quite a stressful one needing not only the shearer but someone to roll and tie the fleece when it's removed from the sheep. When I lived in London I travelled on the Underground sometimes. Advertisements proliferated on the walls and at one time there were a lot for wool and the little ditties always ended up with the line:- "There is no substitute for wool." I remember one of them:

> King Midas turned his goods to gold
> By touching them, or so we're told.
> The shimmer of the royal shirt
> Bedazzled, but the garment hurt.
> The royal vest was just as rigid
> And quite abominably frigid
> The king forgot the golden rule
> "There is no substitute for Wool!"

Dave often decided to put dirty wool around the rhubarb or fruit bushes in the front garden with the idea that it would smother the weeds. It had the undesired effect of killing off the gooseberry bushes but he did it just the same!

The fleeces always feel greasy due to the lanolin present and the only time David's hands are soft is at shearing time. Lanolin is still used in the manufacture of cosmetics as it is easily absorbed. Unfortunately, it picks up all the dust and dirt present in the shearing shed.

Keratin is also present. This is an animal protein and is present in hair, nails, feathers and horn – it is very tasty to the larvae of moths and carpet beetles.

Until the Iron Age when sheep shears came into use, the sheep were either plucked by hand or combed with bronze combs. The secret of wool lies in the structure of its fibres, which are covered with tiny scales all pointing in the same direction like roof tiles overlapping. In one inch of wool fibre there may be 2,000 overlapping shingles, which interlock under pressure, heat and moisture to become felt.

Water is naturally present in wool so it is flame resistant and burns slowly, smouldering and charring but giving off little heat; a wool blanket is an effective way of smothering flames. Wool has good elasticity, giving it a springiness that makes woollen clothing wrinkle resistant when dry. Felt, which is compressed wool, is used for covering piano hammers, winter boots for Soviet policemen, felt tipped pens and polishing wheels. Of course most women will be familiar with felting of a different sort if a favourite jersey is washed at too high a temperature of water. Wool can be bent 20,000 times without breaking compared with silk, which breaks after 1,000 bends.

The Wool Exchange in Bradford was once crammed with traders to buy and sell the commodity which had brought wealth and power to England for 700 years; the Exchange, owing to modern communications, is no longer active.

David, until his retirement, (not that he's ever retired), was a delegate to the British Wool marketing Board and has seen the decline of the great industry, which once thrived. In the House of Lords the Chancellor sits on the 'Woolsack' as a reminder of the origin of England's wealth; many great churches were built as a result of the wealthy wool merchants having the money to do so.

As a delegate he had to attend two meetings a year in Bradford and had the opportunity to go on various interesting 'farm walks' to which wives were invited. As these events lasted a couple of days it made a little break from work for us both and gave the women a chance to dress up for the annual dinner on the evening of the first day. We stayed in a comfortable hotel next to the Wool Board headquarters in Bradford for the A.G.M. One year one of the interesting visits was to the Wilton carpet factory in Bradford.

There are a thousand breeds of sheep- more than a billion sheep graze world pastures. Australia produces a quarter of the world's supply of wool in spite of not having sheep until they arrived from England in the eighteenth century, in boats whose passengers were mainly convicts. The first sheep barely survived the passage and most of them were eaten. Some of us wear 'fleeces', a very handy garment but one that is usually made from man-made fibres. Jerseys once made from wool are now made from acrylic or nylon.

Wool is worth very little now. In 2008 the wool cheque that David received for the wool from his 400 ewes came to £89. This wouldn't have paid for the contract shearers who charge, and deserve £1 per head. Pity the wool was no use in the garden; we could have started up a cottage industry!

46.   Lost Sheep ... and Cats

"HAD A GOOD MORNING DEAR?" I queried, as a good wife should when David came in for his lunch, expecting the usual cheery reply so I was surprised by his answer.

"No, I haven't had a good morning, it's been dreadful. I've lost a ram somewhere at Vernham Dean, thanks to Sid! I've wasted an hour looking for it but it seems to have vanished."

"Surely someone must have seen it; it's not something that's usually somewhere other than a field. What happened?"

"I wanted to move the rams that were in the field next door to the Manor. They needed isolating further away from the ewes as with tupping only a couple of weeks away I don't want them getting ideas too soon. Somehow while I was positioning the transport box, they escaped; I managed to get five of them back but three have disappeared. I eventually found them in an arable field behind the school. I set up the Land Rover and trailer again and secured Sid with a piece of string through his collar. I sent Bud round the opposite way round the rams and started off the other way. Sid had too much energy and he wanted to help, the next thing I knew he had yanked the string from my hand and was tearing round the rams. The three rams went in three different directions and could they move! Eventually two of them were safely in the horsebox with their five companions but there was no sign of the last one. I started looking for it, surely it couldn't have gone far."

As Dave was telling me the story, it reminded me of the story Jesus told in the Bible of the lost sheep, where the shepherd, himself left the ninety nine sheep while He hunted for the one that was lost. In the Bible we are likened to lost sheep when we stray away from God; Jesus is the Good Shepherd who searches for us to bring us back to Himself.

Dave continued, "I warned everyone that I saw to keep a look out for my missing ram. Sheep are funny creatures and don't like being by themselves and become uneasy if they're alone."

While we were having our lunch the 'phone rang.

"I've found your ram Dave, I've put it in with the others, it had got into the garden at the Manor." It was the gardener at Vernham Manor so Dave was able to eat his lamb chops and mint sauce in peace.

Sammy the cat, was suffering the effects of old age and frequently stopped in his tracks to remember where he was going. He liked to try different places to sleep, one week it was the loo seat and he was most annoyed when he had to be removed. We were given a large computer mouse mat by one of the sheep feed firms on which he'd lie in front of the Rayburn as close as he could get. He quite liked sitting in the lane and would be most annoyed if he had to move for traffic but he liked being made a fuss of by any pedestrians.

For most of our married life we've had a cat. We're without one at present as with Pip indoors it wouldn't be fair on either of them. When we had our last memorable cat he got on really well with Jake who was our indoor dog then. When he was a kitten he would cuddle up to him and sleep by him. Paddy (Sullivan) was white and black, very nicely marked and we had him from six weeks old. He led a mainly uneventful life for some years and then one evening he didn't come in. We searched all around but there was no sign of him and nobody had seen him. In the end we had to go to bed and hope he'd come through the cat flap during the night but he was still missing in the morning. As the weeks went on we had to get used to life without him; we didn't get another cat in case Paddy came home.

In the end, after about a year, I decided that he was never going to come home so contacted the Cats Protection League to see if they had any cats needing rehoming. I ended up with Sammy a long haired black cat, quite ugly compared with my Paddy but looks aren't everything and he soon settled down. He was indoors much more than Paddy and had a box in front of the Rayburn. The first time I let him outside I was rewarded by the gift of a frog! He let it go and it started hopping around the kitchen. I fetched the coal shovel and managed to upturn it over the frog. I tipped it up so that it couldn't escape and found a piece of cardboard, slipping it under the coal shovel thus trapping the frog. Sammy seemed puzzled as I think he had brought it in as a present.

The dogs got on quite well with him and he was quite friendly towards them. Then it happened. I had been in all evening and went out into the kitchen to make a drink. There was suddenly the sound of the cat flap – Sammy was in the sitting room. I cautiously opened the utility room door and there he was. Paddy! I quickly blocked off the cat flap and approached him carefully. He looked at me without much recognition. I opened the door into the kitchen and went to tell Dave the good news. We left the doors opened and after a little while Paddy came in looking round carefully. He froze when he saw Sammy asleep in front of the fire, who woke up when Paddy started investigating him with his nose. They both had a good sniff of each other and then settled down to wash. I don't know if you know the adage that applies beautifully to cats, "When in doubt, wash!"

Paddy settled down again with us and an uneasy truce between the two cats took place. The evenings were particularly awkward as they both laid claim to my lap. I often found myself sitting on the settee with one cat on my lap and the other draped along my shoulder – I felt like Long John Silver! They each had to have a dish for their food and Paddy liked to sleep on the armchair while Sammy had a cardboard box in front of the Rayburn. Things carried on like this for several months and then Paddy started staying out at night. He'd stay in for a few nights then miss some. Once again David started scouring the countryside and eventually discovered that he had been seen at a farm a mile away. By the sound of things it was him as he was quite distinguished with his mainly white coat and black tail. He was catching the vermin in the cowshed and the farmer didn't mind him there at all.

We felt that it was pointless trying to keep him at home and as I still had a cat to sit on my lap I was quite happy. We had a cold winter that year and I discovered that Paddy was spending time in the house of the farmer's daughter. He'd go in during the evening and curl up on the dogs' bed for a sleep. Then he went missing again and nobody saw him after that. We would love to know where he went. We heard that he had also been spending time at another house half a mile away and the people had moved about the time that he went missing. We'll never know if he ended up in a town being called 'Patch'.

After Sammy died we didn't have another cat as I had 'Woll' a ten year old retired sheep dog. David liked to retire his dogs at ten so they then became house dogs and had to learn a whole lot of new skills. It was hard for him to get used to walking on a lead as until we reached the field we were on the road. He wasn't keen on the fire and moved away as far as he could. One thing all the 'house dogs' found out was that if they were in the kitchen when I was preparing meals, bits might fall on the floor – they were particularly fond of sultanas.

47. Dave's New Hips

DAVID WAS BECOMING AWARE that his hips were beginning to be painful, especially if he had been doing a lot of walking. He went to visit his doctor, who sent him to see the appropriate consultant after having X-rays of both hips. It was discovered that he had arthritis in both hips and would need replacements, one at a time. It was decided to do the first one after lambing so he was booked to go into Basingstoke Hospital, which is about three quarters of an hour's journey away. He had his left hip replaced and the next day was walking around with a crutch and came home on the fifth day.

He had quite enjoyed his stay but was determined to return home as soon as he could so was meticulous in doing all the exercises he was given to do. He had been surprised to notice that some of the other patients were reluctant to try anything and were content just to lie in bed. He continued his regime at home and found that Horseshoe Lane was just the right length for a walk so had at least two per day. He had had to have some minor adjustments to furniture or equipment such as a grab rail by the shower and blocks under the legs of the bed. He soon became fit again and was discharged from hospital care at his three month check-up. The second hip was done a year later; he rolls a bit when he's walking, rather like a drunken sailor but six years on he's had no more bother.

48. More About Dogs

OUR LAST HOUSE DOG WAS PIP, I've written about some of his earlier exploits elsewhere. He lived to the ripe old age of thirteen and enjoyed his life as a house dog .He went to work in the morning with Dave and the idea was that Dave (sometimes) finished work at lunch time and pottered in his garage/workshop in the afternoons. He's rebuilding an old Massey Ferguson tractor and at last seems to have more time to spend on the project. Pip and I went out for a walk at some point in the afternoon although it was never very far as there were too many interesting scents around and he had to investigate every one. He found a flattened rabbit head once and I had to prise his teeth open to get it away from him.

He enjoyed 'helping' me when I was cooking, especially Saturdays when I do a big bake for the following week; he sat at my feet and it was hard not to fall over him. In the evening he joined us in the sitting room starting off on his blanket but moving around the room whenever he turned over. He was very partial to apple cores and I discovered that he liked raw carrot also; he'd eat chocolate if we let him but we knew it isn't good for dogs. He loved coming out in the car and sat on a rug on the back seat looking out of the window or else just slept.

He had an infuriating habit of howling if he was left by himself, sometimes even indoors if he couldn't find one of us. He'd start off by clearing his throat and, I think, selecting in which key he was going to sing. Then his head would go back and he'd 'sing' one chorus, finish that and start on the next. When we left him in the car he'd sing a couple of verses then settle down to sleep. It's odd that these old sheepdogs that have never been in the house make themselves so much at home when they retire.

David has two other sheepdogs; Bud who is seven and Sid who is two. They work very well together and are first rate dogs. Bud has a habit of breaking out of his kennel especially when there are fireworks around. Thankfully, he's well known in the village and we'll get a 'phone call to say that he's been found and can we go and collect him. The Land Rover doubles up as a mobile dog kennel as Dave has a wire grid at the back so that the dogs are safe. Bobby, one of our previous dogs fell out and was run over before Dave had the cage. Pip disdained the back – it was a bit of a squash in any case – and travelled on the front seat, he couldn't jump up any more so Dave had to lift him up. I also get lifted into the Land Rover on the odd occasions when I travel in it as I can no longer climb into it!

It's not often that I need to travel in the Land Rover but sometimes I go out with Dave on Sunday afternoons when he takes the dogs out .We often go to an area of Forestry Commission woodland at the top of Windmill Hill. There are several walks there and it is interesting when the snow is on the ground to see the deer and bird tracks crisscrossing the woods. The dogs love it up there as there are so many new scents and they enjoy rolling in the snow.

Another time we'll drive along the tracks that cross the area, bordering the fields. Dave lets the dogs out and they run behind the Land Rover running through all the puddles. They're really quite handsome with their black and white markings and a white tip at the end of their tails but they often become a uniform mud colour, which, incredibly by the next morning will be cleaned off. Sid likes stopping to investigate the Kune Kune pigs that are snuffling in a wood.

Pip became slower and slower and was reluctant to come out for a walk. He had difficulty standing up and became incontinent so we felt that the time had come to have him 'put to sleep'. We do miss him a lot but knew that he had enjoyed his life.

49. My Parents' Last Years

MY SISTER DIED OF CANCER when she was fifty eight and my parents decided to move nearer to us. I managed to arrange for a transfer for them to Andover and found a nice flat with a resident warden, which we all felt that they needed. They liked it at first, but, as usual soon found that there were drawbacks to living in the flat as nowhere is perfect. There was a laundry room with a tumble dryer, which Mum refused to use so did her washing in her sink and draped it around the flat to dry. She was able to walk to the town centre and I'd take her to get her main shopping once a week.

They lived there for about nine months when Dad died of a stroke after being in hospital for some weeks. I thought Mum would want to stay in the flat as it was so convenient but she decided she wanted to move back to Telford! Unfortunately, she exaggerated the snags about the flat to my brother in law and he managed to persuade the council to give her a flat in Telford. Mum was only there six weeks when I had a pathetic letter from her acknowledging the fact that she had made a mistake! The flat was on a different estate from my brother in law so she didn't see much of him, there were a lot of foreigners around, she was frightened of the dogs that roamed the estate and please could she come back to Andover!

I also had a letter from Wally as he realized too late that she had this problem of thinking that 'the grass was always greener on the other side of the fence.' By this time she was in her nineties but I felt that I'd have to try and get her back in the locality but this time she'd have to stay put. I applied to the council once again and eventually they decided that as I was her only close relative and my sister lives in Australia they would put her on the list. After a few weeks they 'phoned to tell me that there was a bungalow in our village available, would I like to see it? It was only ten minutes' walk from us and it seemed ideal but I decided that she must see it before taking it.

We met my brother in law halfway between us at Stratford-on-Avon and Mum stayed with us overnight. I took her to see the bungalow; it was a one bedded semi-detached place with a nice garden at the front and grass at the side and back. It was up quite a long path but it was ideal inside with a kitchen, bathroom, bedroom and living room. She loved it so we went into Andover to the council offices to arrange everything. She had been at Telford just ten weeks!

Mum really settled down in the bungalow (she knew there was no chance of moving again) and in the afternoons would sit out in the garden with her knitting; George in the adjoining bungalow would be in his garden with his book. They never spoke much but Mum felt at ease with him just being there. She was even more pleased to discover that he enjoyed watching snooker on the television. She watched nearly every game and so did George; they enjoyed discussing the matches and the players the next time they met. Mum's favourite player was Steve Davis. The person she really disliked was Stephen Hendry so if the two of them were playing each other you can guess who she was cheering on. George was fond of darts too but Mum could never discover what the rules were.

She could walk to the Methodist chapel and went to the service there on Sunday mornings and Tuesday afternoons to the Women's Fellowship. There were a couple of ladies from the other bungalows who went so she had somebody to walk with. On Sundays she'd come to us for lunch, which she always enjoyed and I'd take her back afterwards.

Mum loved coming to town to do her shopping; Dora came with me as well and we would meet up with Ted and Lily for 'elevenses' being met also by another couple of friends. We had our break in the Guildhall where a local charity had the franchise to sell refreshments and have a couple of stalls. I'm afraid we were rather a noisy crowd but nobody seemed to mind.

Although Mum never became a member she was invited to the outings of the over 60's club, which met fortnightly in the British Legion hut. I went along as well and we had some memorable outings to Bournemouth and Poole – where I had the job of finding two ladies who had failed to turn up for the coach home. They were so grateful when I found them as in the big shopping precinct all the aisles looked the same and they had no idea how to find the way out.

Mum was in the bungalow for about four years but had to go into full time care when she broke her hip. I had had to take her to Salisbury Hospital to have a cancerous blemish removed from her face. It was a rough windy day and I didn't know the layout of the hospital so parked too far away. It was hard to make any progress against the wind and I had to ask someone to help me walk Mum along. She had the growth removed and didn't seem too worried by it. She coped all right but had a fall after she arrived at her bungalow when I went home to see to things there.

She was able to activate the alarm system and I was contacted to go and see what had happened. She was on the floor but the ambulance was soon there and she was taken to Winchester hospital where she was diagnosed with a broken hip. It was the first time that she had ever stayed in hospital. Although the break mended all right, it was obvious that she needed full time care as she was about ninety five and now needed a walking frame. She was admitted to Cherry Orchard Retirement Home in Andover where she lived until she needed more care so went into hospital where she died after a short time at the age of ninety nine!

She had been the eldest of six children and had an odd childhood as her Mother died at a fairly young age. The lady her father subsequently married was Dad's mother who had four children so there was a problem about finding a house big enough for them all. My surviving aunt who was the youngest of the six children remembers that she had to go and live with a relation. Mum had to sleep in the kitchen on the 'settle' until a camp bed was found for her. She went to work near St. Pauls Cathedral at a shirt making factory but eventually married Dad and they moved out of the family home.

My sister Pam and her husband emigrated to Australia under the £10 scheme back in the late fifties. Her husband had been in the army; one of her children was born in Cyprus and the other in France. She's only been back once since she's been out there, Dave stayed with her when he went to the Antipodes with his Nuffield Scholarship but we keep in touch with letters and the occasional 'phone call. My other sister, Joyce, married and lived in Telford, she just had one son, Michael and died of cancer when she was fifty eight.

50. Ducklings

WE'VE ALWAYS MANAGED TO GET AWAY for a holiday in the summer and for some years have been traveling to different places in France. Our knowledge of French is minimal but we can manage to hold a reasonable conversation. One of our visits was to the Loire valley where we enjoyed visiting the various chateaux but didn't see many sheep! We did, however, find a village agricultural show, which was most enjoyable with all the various animals – sheep, cows, poultry and mules – the equine sort, which could be heard all over the show ground with their hee-hawing. Tractor rides were popular and there was a display of log cutting. Marquees with tables and chairs were provided, in which people could sit and eat their own packed lunches. It was all very French and enjoyable. As we passed through the village, we saw that all the houses were decorated with fruit, flowers, and vegetables with more than one house sporting a scarecrow in a wheelbarrow outside.

It was soon back to work for Dave as the sheep always need attention. It's quite a problem keeping them fit and healthy; a lot of the work is preventative measures as parasites are always present, both internal and external.

Ticks, mites and lice are external parasites that live in a sheep's wool. They are not only an irritant to the ewes and lambs but can also cause disease in the animal. Ticks feed on the blood of the sheep but can also live on other animals, even humans! If a tick is pulled off it leaves its head behind attached to its host, which goes on feeding. The treatment for this is similar to that used for blowfly strike, which is a pour on chemical. Unfortunately, if humans get a tick bite it can set up an infection called 'Lyme's Disease', which is a very debilitating illness and can result in death. It is very prevalent among deer herds and people who work in the woods have to be very careful. The ewes used to be dipped in the summer but it is now permissible to use 'pour on' or 'jetting' procedures.

I hope this won't stop you eating lamb, which incidentally is not used for meat until the lambs are 12-18 months old and not the little woolly things one sees in the meadows; without the demand for lamb meat there wouldn't be any lambs and David would be out of a job. The wool they produce is worth very little now as the various man-made fibres have taken the place of wool. Wool needs a lot of processing to get it from the sheep to the finished product. Shearing is a very demanding and skilful job and best done by younger men as there's a lot of bending and manoeuvring of the sheep. David has taught innumerable people to shear and whereas at one time it was the preserve of the men, now there are excellent women shearers.

We haven't had any ducks for a few years now; Mr Trewby's disappeared at the same time, I wonder if somebody fancied duck à l'orange for dinner one evening? We had two ducks sitting on eggs underneath the shed, Rachel and Leah. Rachel's eggs hatched, there were nine ducklings but one died in the first week of its life. Leah was convinced they were hers but as she was only sitting for a few days it wasn't possible. Rachel was very generous though and let her share the sitting with her. Ducklings are quite self-sufficient and manage chicken pellets from the beginning. One day they found an ants' nest and found them very tasty. They waddled along on their minute webbed feet, which are really redundant as Muscovies don't like swimming. I had to live in a state of siege for a few weeks and keep the outside doors shut. They discovered the route to the cat's food at a very early age; if they get in through the front door they have to traverse the carpet – and they're not house trained! They'll stand in the cat's milk and devour his cat food at an alarming rate, I then have the task of removing nine ducklings out of the house, flapping a towel at them seems to be the best way. It's a good thing that we have patient neighbours as the duckling love getting through the fence to pastures new. The two mums get very agitated until someone appears to open the gate and re-unite them all. Woe betide anyone who tries to pick up one of these fluffy yellow toys, Dave's boots still bear the marks of the ducks bills where they've attacked him.

As the ducklings grew it became increasingly difficult to differentiate between Rachel and Leah. At first the red part of Rachel's face was quite pale probably due to sitting in the dark under the shed for the month it took to incubate her eggs. It then turned bright red like Leah's and even the markings around the ducks eyes looked identical. The sisters huddled together to sleep but the ducklings only crept beneath Rachel, they seemed to think that if their heads can't be seen that they're invisible.

The dogs were frustrated because ducks don't normally react like dogs; I think they heard about a dog and duck demonstration at a Flower Show. Sammy the cat just ignored them completely and stared disdainfully at them if they tried to go near him. Danny, the father of the ducklings wanted nothing to do with them. He kept Bertha the bantam company but one day Dave brought him a lamb as a new friend. The lamb was quite poorly but after treatment Dave thought it might recover with some TLC. It was unable to walk but with fresh green grass and sheep nuts to eat, plus a bowl of water within reach it seemed quite contented except for one thing – flies. They would not leave the poor lamb alone; but catching flies was one of Danny's skills. He sat next to the lamb all day catching any flies that might land on it. He seemed to be saying, "There are no flies on her!"

51.   A Bad Cut for Dave

DAVID IS A TRUSTEE OF HAMPSHIRE CHRISTIAN TRUST, which is a trust set up to manage a camp site at Lockerley for Christian camps. The Trust look after the amenities at the site and during the summer, church groups can use the facilities. At first there was just the field to manage as the campers slept in tents; but as the years have gone on several buildings have been added. Now there are two large barns, which house a well-equipped state of the art kitchen, toilets and showers, and a large meeting room plus an upstairs 'quiet room'. There are toilet and shower blocks outside and the waste disposal is taken care of by a 'reed bed' system.

The site is about half a mile from the road in a secluded field surrounded by woodland. The local community uses the barn at Christmas for a village carol service, which is always well attended. Although we've never been able to take an active part in camp work owing to the nature of Dave's job, we have always been interested in it, and know that many young people have committed their lives to Christ while at camp. There are camps in progress right through the summer from spring Bank Holiday onwards.

One December David and Andrew decided to do some maintenance work with some other trustees on the camp site. Their job was to demolish an ex-British rail bunkhouse that was no longer needed on site. Andrew was on top of the bunkhouse while David was working on the side panel with an angle grinder, they wanted to save the aluminium to take to Joe Hirst the local scrap dealer. Suddenly Dave let out a shout and the angle grinder fell to the ground.

"Andrew! I've cut my arm"'"

Andrew scrambled down from his perch. Blood was covering David's wrist.

"There's a first aid kit in the front of the Land Rover, hold your arm above your head and it will help to stop the bleeding." Andrew advised.

He soon was back with the first aid tin and picked out the biggest dressing, covering the gaping wound with it. "I'll just tell the others what's happened and then I'll take you to Salisbury Hospital."

Andrew covered the twenty miles in twenty minutes and parked in the ambulance bay!

"It's an emergency," he explained to an official and by now the bandage round Dave's arm was soaked through with blood. Andrew found the way to the A&E department and soon a pretty nurse was attending to Dave. As she unwound the bandage there was a sudden spurt of blood, which if she hadn't been wearing a plastic apron would have made more of a mess than it did. She hastily put a bigger dressing on the wound and sent for a doctor. His opinion was that Dave had nicked a blood vessel, which needed stitching and this would have to be done by the 'plastic' surgeon so Dave would have to stay in overnight. As there was nothing else Andrew could do he went home and phoned me to tell me the news.

An hour or so later Dave used his mobile to phone me from the hospital as he was allowed to use it. He wasn't able to have the procedure until Sunday, the next day, and would have to be fitted into the surgeons list so might not be home until Monday. He had nothing with him; no money, glasses, clothes, or book, was in a room by himself, hungry and bored and the television wouldn't work! His arm was suspended above his head in a sort of foam rubber sling. A friend, Anita, kindly took me to visit him and so I was able to take his glasses, book, pyjamas and toiletries and Anita managed to get his T.V. working The nurse brought him a big plate of pasta , the last meal he had eaten had been breakfast. Because his arm would have to be dealt with under anaesthetic, he wasn't allowed anything to eat the next day and had to wait until the surgeon had a slot for him.

He didn't go to theatre until 4.30.p.m. having been allowed one slice of toast at 6a.m. and finally had a meal at 9p.m. I collected him the next day and we had lunch in the hospital restaurant before setting off for home.

Rumours had been rife in the village and more than one person thought he had lost his arm. I think he'll be more careful in future. There was an article in the paper that an angle grinder can be used to destroy a computer!

52.   A TV Appearance

DAVID IS GRADUALLY CUTTING THE FLOCK NUMBERS down and plans to sell another forty ewes after weaning. He just rents some fields in the immediate locality to graze the two hundred ewes and their lambs, these, of course will be sold for meat in a few months' time. He rents other fields for the grass on them so that he can make hay. It's been a difficult job this year and sheep feed is going to be short. The hay that he cut earlier in the year was very light and a field that yielded 900 bales last year only produced 400 bales this year. Unfortunately they couldn't be brought in soon enough and got wet so some were useless. Hay 'sweats' and can spontaneously combust if it's stacked too early. Dave's cutting hay today hoping that the forecasters have got it right. Sheep also eat a proprietary sheep food in the winter but this can prove rather expensive. He's been offered a big field of grass that hasn't been grazed for some time so that will help to tide us over.

Since 2007 I have had to have frequent visits to the Eye Clinic at Southampton Hospital as I have Age Related Macular Degeneration. In my left eye it's the 'dry' variety for which no treatment is available and it's slow developing. The right eye has the 'wet' type and there is a treatment for it, which necessitates an injection into the eye itself. Until 2008 the procedure was unavailable on the N.H.S. so we decided to pay for it as without it I would lose the sight in that eye. Relations paid for the first one, for which we had to go to a clinic near Portsmouth. It wasn't a pleasant experience but the eye was anaesthetized so there was no pain and the eye was held open with a special instrument so that I didn't blink. Three months later the procedure was repeated at a hospital at Chandler's Ford, which we funded ourselves.

Then we heard that The N.H.S. had decided to fund it. We arrived at the hospital as I needed another injection, and soon had a sight test, the first procedure, followed by drops being put into my eyes. David, having had difficulty parking the car, joined me and we were sent into the next waiting room. After waiting for what seemed to be hours but in reality was about ten minutes, the summons came that the doctor was ready.

"Can my husband come with me please?"

"Yes, that will be fine."

We followed the nurse into a small room, which seemed to be full of people. "Just sit there." I sat where she indicated and a female doctor joined us. Dave made himself as small as possible in the corner of the room.

"Mrs Sullivan would you mind if the procedure is filmed as you are one of the first to have this done under the National Health Service and Meridian want to film it for their news programme?"

I glanced at Dave and he nodded, I thought that it might help somebody so why not, as long as I wasn't asked any awkward questions. The first thing the 'crew', consisting of a sound man, camera man and presenter wanted to film was me signing a consent form. Then I had to lie down on the couch where the doctor explained the exact procedure; the presenter then stepped in and asked if I was pleased that it was available under the N.H.S. now. My reply was that I had just been glad that treatment was available. The doctor and nurse came closer and were checking things just out of my range of vision. I had two more lots of eye drops administered then a small 'sheet' with a hole in it was draped over my face with just a square for access to my eye. Just then, the camera man stood up, he had been trying to get a different angle. 'Clang!' his head collided with a piece of static machinery and he staggered around holding his head and groaning! He eventually stopped, "Sorry, I'm all right now."

"Could my husband hold my hand please?" Dave moved in and grabbed my hand; he had to spend the next few minutes kneeling on the floor as there was nowhere for him to sit.

"I'm going to clean round your eye. This retractor will keep your eye open. There that's done it. Syringe please nurse." The camera man moved round to get the best shot. The Doctor raised the syringe; I squeezed Dave's hand and also the nurse's. The needle went into my eye, it was uncomfortable but not painful, but the T.V. crew wanted more.

"Doctor, could you just stand in the same position with your hands over Mrs Sullivan as you were for the injection?" The camera man and the sound engineer moved into position again.

"That's fine. Mrs Sullivan, I'd like to ask you a few questions."

I hoped this would never go on air. I found myself saying such trite things as I was asked obviously leading questions about the injection I had just received and also the two that I had received privately. Eventually they were finished and the doctor was giving me various instructions. In the background I could hear the presenter talking to the camera man, "Stop making such a fuss; I know you've got a bruise but be thankful you haven't had what Mrs Sullivan's just had!"

They wanted a final shot of my hand being shaken by the doctor and of me walking out of the door.

Meridian News did screen the item but unfortunately we come into the Thames region so didn't see it. Andrew managed to find it on the Internet – it lasted one and a half minutes!

This year I had an injection in January then went right through until October before I needed another one, having had several check-ups in between.

53. A Trip to Belgium

EARLIER THIS YEAR we decided to have a mini break and settled on a Eurostar trip to Bruges, which we found very interesting. We very rarely go anywhere by train so this was a big adventure for us. Our neighbour took us to Andover Station and we were off! We decided to have a taxi across to St. Pancras so saw a bit of London on the way. The station at St. Pancras resembled the inside of a cathedral with its soaring roof and feeling of space. We checked in and soon were comfortably ensconced in a carriage ready to go.

We seemed to travel through the London suburbs quite slowly at first but soon were speeding along until we reached Ebbsfleet where the train stopped to take on more passengers. After that it wasn't long before we were under the channel; quite an eerie feeling really to think of all that water on top of us. We soon came to the surface again and in no time at all we were at Lille where we alighted. Here, we transferred to coaches for the last part of the journey, which took about an hour. The courier in charge of the coach gave a running commentary about the countryside through which we were passing and soon we reached Bruges.

It's a very old town with cobbled, narrow streets and not much room for a coach. It parked outside our hotel and it was quite a scramble to get in as the coach was blocking the road. We had a very big room with even a hospitality tray, which isn't usual on the continent; we had to buy our own milk though, which David was able to buy from the supermarket next door. It was early evening by now and the courier had already given us information as to where the restaurants were. After a cup of tea we set out to explore. All the important buildings were floodlit and there seemed an abundance of restaurants so we picked one at random and enjoyed our first meal in Belgium.

The next morning the guide took us on a walking tour of the town and we also had a tour of a brewery with many steps to climb. The morning ended up with a visit to a chocolatier and most of us ended up in the adjoining restaurant for lunch. In the afternoon Dave and I went round an interesting museum and browsed round the shops. The next day we were loaded up into the coaches again and were driven to Ypres via a big war cemetery. It was very peaceful there but we realized more than ever what a terrible thing war is. The graves stretched away into the distance.

We went round a museum in Ypres; the Continental people have a real flair for displays of any kind and this was no exception. We had a lovely meal and it was soon time to board the coaches again ready to return to the hotel. The next morning we had to pack up although we were not leaving until after lunch. A boat trip on the canals was planned but unfortunately, I was unwell so we didn't go. We left to go home and finally arrived back in Andover at 9 p.m. after a most enjoyable three days away.

54. The Dangers of Geese

WE ONLY HAVE A SMALL YARD at the back of our house, which
contains the dog kennels. David made these himself, they are
raised a couple of feet from the ground and consist of a slatted
floored run with a kennel at the back and a roof covering the
whole edifice. Sundry birds or animals are moved into the
paddock behind, of which we have the use from time to time
and I had noticed three geese at the back of the field.

I had just hung some washing on the line when I became
aware of the geese approaching. I started to back away from
them, thinking that it wouldn't be wise to turn my back on
them; I grabbed a thick stick from the ground as I went. I
glanced behind me to make sure that I was going in the direction of the gate, which was about four yards away. Suddenly I
found myself falling backwards and 'bang!'. I fell heavily hitting
my head on something as I fell. It was the water trough!

My head hurt. I put my hand up to it and it came away covered in blood. I realized that I'd have to get up and knew that
both lots of neighbours were out. I was still clutching the stick
and by leaning on it prised myself off the ground and staggered
indoors. I reached the 'phone, stopping to snatch up a towel as I
went. I tried Dave's mobile – no reply. I tried a neighbour – no
answer. I tried a friend who lived further away although I knew
she couldn't help but she may have seen Dave. She was in and
said she'd phone her daughter in law, who was working fairly
close by.

Jane was soon with me and tried to staunch the blood deciding that it was only a flesh wound. She had just succeeded when Dave arrived. He prised the 'cudgel' from my hand, I hadn't realized that I was still clutching it and he decided to take me to the Minor Injuries unit at Andover Hospital. I was soon attended to and came away with the edges of the wound glued together. I was left with a very stiff neck and shoulders and was pleased when Dave suggested that I should make use of an Osteopath who had opened up a practice in the village. I had three sessions and was relieved to get rid of the stiffness.

I was reminded of a run in that I had had with a different animal a few years ago. I was hanging out the washing – I do other jobs too – and having finished, I became aware of movement behind me. I turned to see and was promptly butted to the ground, landing in a clump of stinging nettles. The culprit was a three legged ram that was living in the paddock temporarily. I shouted at it struggling to my feet and picked up the laundry basket to throw at it. Somehow I managed to get across the paddock to the gate and safety. Apart from being covered in nettle stings only my pride was hurt.

Andrea is just coming to the end of a two year course at Sparsholt on Animal Management and would like to work with dogs. Bethany starts at Bournemouth University in September and wants to be an Occupational Therapist. Daniel is just sorting out his GCSE options and Michael is doing well at school and enjoying cooking for everyone. He won an award the first year he was at Senior school for 'Continued Effort' and had a silver cup for a year. His confidence has increased tremendously and he is a patrol leader in the local Scout troop.

55.   Another Setback

DAVID HAD A COUGH, which I had never known him have before. I bought a bottle of cough mixture and managed to dose him up with it. It became so persistent that when he was speaking, especially in public he was coughing continually. With a lot of persuasion he went to his G.P. who gave him a course of steroids to take. The course lasted a fortnight and still he coughed. He went back to the surgery and saw a different doctor, he received antibiotics this time, but after completing the course, and then a second course, he was still coughing so an appointment was made for him to see the chest consultant. He had an X-ray, CT-scan and finally two needle biopsies before it was finally diagnosed that he had mesothelioma in the lining of the right lung. This cancer is nearly always connected with contact with asbestos at some time in the past although Dave can't remember any specific occasion when he was involved with any. It was decided to start chemotherapy.

This necessitated a trip to Winchester Hospital every three weeks, usually arriving by nine o'clock where he'd spend the next three hours hooked up to a machine which administered drugs via a needle in the back of his hand. I went with him armed with books, snacks and TLC. Everyone was very cheerful and there would be another four patients also having chemotherapy in the room, all sitting in outsized armchairs. Armchairs were also provided for escorts so it was a good opportunity to catch up on some reading. Drinks were available the whole time and lunch could be ordered if one was going to be there for that length of time. We usually went into Winchester afterwards and had a meal out; we found a good Cornish Pasty restaurant that we liked.

The first three treatments he had didn't seem to affect him at all. They didn't get rid of the cough either and he was feeling quite rough and cancelled all his speaking engagements for the next three months. His appetite disappeared, he started losing weight, and he just slept a lot. He managed to keep up with his work on the farm although at that time of year there wasn't a lot to do once he had sold the lambs. The next two courses of treatment left him completely shattered, he couldn't think straight or concentrate on anything. I tried different meals to try and encourage his appetite, swapped sunflower spread with butter and full cream milk in his tea as well as on his cereal. He was offered a place on a trial for a new drug that was being investigated worldwide to think about. Christmas shut everything down but David began to start feeling better. He sorted out some more scrap metal to go to Joe Hirst, the scrap metal merchant, and was pleased to discover it was worth even more than the previous trailer load he had delivered.

He read carefully the eleven A4 pages outlining what the drug treatment involved for the trial. The possible side effects were horrendous, the trial was over two years but then he read that the trial was a 'double blind' one. He wouldn't necessarily get the drug but might receive a placebo.

He was beginning to feel better, the cough was receding, his appetite was returning, he was gaining weight, and he was feeling generally more able to cope so decided to refuse the offer of the drug trial. The specialist who was dealing with the applicants agreed with Dave that it would be foolish to jeopardize the quality of life that he had and said that he wasn't 'desperate' enough to take up the offer. David has put back the weight that he lost while undergoing chemotherapy and feels really well again with no trace of a cough.

Since then he has continued to improve and in another three weeks will be lambing his hundred and eighty ewes. He sold his eight barreners for more money than he paid for them eight years ago! He has decided to sell his ewe flock at Wilton in the Autumn and have ewe lambs on keep for other people. This will do away with lambing – although he will keep Bridget and some of the other odd sheep. He still has grass available for sheep and has already spoken to friends in Kent who might be interested in his scheme.

56. Music and Talks

DAVID HAS RECEIVED A LOT OF PLEASURE from trying to learn to play the banjo since he's had more spare time in the afternoons. He had started making one himself at Evening Classes some years ago but the tutor died when he had made the bowl – using my biggest saucepan as a mould and bending the different layers of wood round it- and also the neck. He had no idea how to string it and it stayed untouched in the box room for the past twenty years. He was booked to speak at a harvest supper at a church in Cadnam last year and they asked him to join in with their music group for the evening as they had been asked to perform. At that time he played the ukulele, not very well but enthusiastically and felt privileged to be asked.

He had taken up the ukulele when the boys were small as they had wanted guitars to play, which were much more expensive but as the boys were only five and seven it wasn't worth buying any. I discovered ukuleles in a music shop for two pounds each and they were quite happy with them. Dave picked one of them up and discovered that, although it was only a small instrument, the four strings were further apart than on a guitar, and his big, stubby fingers could manage them much better than a 'proper' guitar.

Over the years he's enjoyed using it at various meetings and is particularly fond of playing 'Box Car Willie's,' "I wake up every morning with a smile on my face!" He thoroughly enjoyed his evening at Cadnam, and happened to mention that he had a half-finished banjo at home. One of the group offered to finish it for him and did just that, stringing it, tuning it and adjusting it – it's a beautiful instrument now and David is on the way to mastering it. He has obtained a book 'How to play the banjo' and spends some evenings in the kitchen with his beloved instrument. The snag now is that he has arthritis in his hands and can't manage the fingering very well.

He goes to various places to give talks apart from the many churches that invite him to speak. Some of the venues will only have a small audience while others – particularly the 'U3A's will have a hundred or so. 'U3A' is an acronym for 'University of the Third Age' and there are branches all over the country. It's been set up by pensioners and a group will discover what abilities are available within the group and those that feel able tutor those that want to do the respective skills. So a group might consist of a French teacher, Gardener, Pianist, and Computer buff who will arrange lessons in their respective subjects. Once a quarter, the whole group will meet together and a guest speaker booked, which is where David comes in. One group in Hampshire numbers over one hundred.

This afternoon he is speaking at a group in Andover known as the 'Homemakers', they discuss problems in the home, ex-change recipes and household tips. They finish with a cup of tea and exchange any thing they've brought to trade.

Recently he had a booking at Southampton University speak-ing to their Retirement Association. David's sister from Australia was staying with us and was intrigued to know just what happened at these meetings. The meeting was held in one of the lecture halls with about fifty people in the audience. After some preliminary group business, the floor was David's. He kept the group enthralled for an hour as he told them of his intro-duction to the world of farming on Exmoor. One thing that had impressed him was at harvest time how the neighbouring farmers – although they came from miles away – all arrived to help. He was even more impressed with their wives turning up to help with the enormous tea that followed with everyone crammed into the farmhouse kitchen. Just to sit there and listen to the stories that were told about happenings during the day was incredible to a town boy.

Dave told his audience about the milestones of his life, of learning to shear the 'Bowen' method, of his Nuffield trip to the Antipodes and many other experiences. He spoke for an hour and then had time for questions. Maureen was quite impressed and enjoyed helping me sell copies of my first book, 'The Sheep's in the Meadow – Hopefully'. We went down to Hythe afterwards in the hope of seeing any liners. There were not any in but we had a fish and chip tea, which Maureen had wanted, as there aren't any fish and chip shops in Australia.

David's church engagements are mainly on Sundays but cover a broad spectrum of churches within a radius of thirty miles although he will travel further if it's a special occasion.

57. Friday Club

AS THERE IS SUCH A SMALL CONGREGATION at the Gospel Hall at Vernham Dean that we attend, we are always there on Sunday evenings. In the mornings Dave often has a preaching engagement at another church and once each month we have a Family Service on the Sunday morning, which is very well attended by parents and children from the Friday Club we hold each Friday evening. Ten to fifteen children attend this Club, which is held for the junior school age group and we've been running it for so long that a few of the children have parents who attended as children themselves.

We run it as an old fashioned Sunday School with singing, a 'news time,' with a big cheer if anyone has done well at school and named as an 'achiever'. We then have a prayer by Dave followed by any child that wants to pray. Then a quiz with a game – magnetic darts is a big favourite, as well as an outsized version of 'beat the bleep'. Our son Peter made this out of old copper piping welded together with devious twists and turns and powered by a torch battery. The aim is to get a metal noose from one end to the other, which if it touches the pipe pathway strikes a bell – loudly. The children love it and we often use it at the Family Service when the parents have to have a go.

We have a drink and a biscuit, usually a Penguin and have to suffer all the 'penguin' jokes. David tells them a Bible story and the last ten minutes we divide the children into age groups to complete an activity sheet based on the Bible story. The children are not angels but are quite well behaved and enjoy coming. They all go to the same school in the village, David is a governor, so often sees them when there's a governors meeting at the school.

For the past three years a camp has been held at Linkenholt, a village about three miles from Vernham Dean where the trustees that administer the site have set up camping facilities for local organizations. It used to be part of the Linkenholt estate, which was a farming enterprise but the grazing was rented out and the buildings used for other purposes. One of the Dads arranges the weekend with Dave's help as Friday Club is the 'umbrella' organization.

The camp site is miles from anywhere along a lane and a building has been erected for use as a dining hall and meeting room. There are basic kitchen facilities, a couple of barbeques, showers and toilet blocks. The camping weekend works on the principle that each family does its own cooking, has its own tent and is responsible for its own children! No cooking is done on the first evening as an Indian Take-away is situated in a village about six miles away and meals are pre ordered and delivered to the camp site, it works very well.

The first year a girl fell out of a tree and fractured her arm on the first evening! She went to the local hospital and had it set and came back to the camp to finish the weekend. All ages are represented from toddlers to OAPs and people just take part in what they want to, Dave and I have slept at home for the past two years, we can be there in fifteen minutes. A walk is arranged for the Saturday afternoon, which surprisingly is well supported. In the evening a campfire is lit and silly songs are sung and stories told, the smell of wood smoke is very evocative.

This year a water slide was constructed down the side of a hill with a large polythene sheet, water supplied in a water bowser and washing up liquid. Parents were dispatched to collect swimming costumes from home and the fun began. As the afternoon went by, a few parents couldn't resist the opportunity to relive their childhoods and joined in too. Sunday morning a service is held in the dining hall and everyone turns up for singing of a different sort and a short talk by David. Everybody goes home after that but some of the men return in the afternoon to tidy up the site; it's a really good weekend and as everyone knows each other it's very safe.

Soon after we were married, an aunt of David's came to visit us. She was appalled at the remoteness of our house and the fact that we had no family or friends around.

"My dear, whatever will you find to do with yourself all day?" I find that there's never enough hours in the day to do everything that I want to do and there's always something going on.

Once a fortnight – soon to be once a month because of cut backs – the Library van calls in the village. Ours is the first stop at half past nine until ten minutes to ten when it moves off to visit two more locations in the village. The attendant is very helpful and will get any book for us, usually at the next visit. There's a lot of non-fiction books on various subjects and fiction of every genre. We can retain them for as long as we like within reason and there's no ban on talking in the Library. We chat about any good books that we've brought back and others that we haven't enjoyed; I take an elderly friend in the car as she can no longer manage the walk, nor can I really as I always come away with a pile of books. I go back with her for a cup of coffee and my neighbour who has also been to the library joins us.

Various coffee mornings take place and once a month there's a ploughman's lunch held at various homes throughout the village. Shrove Tuesday is celebrated with a 'pancake morning' when we enjoy pancakes, lemon and sugar plus good company.

Our end of the village is actually a separate hamlet, Ibthorpe, and comprises mainly of a vaguely horseshoe shaped lane containing about twenty eight houses. We feel privileged to live here as it's a very special place.

58. A Golden Weekend

RECENTLY WE CELEBRATED our Golden Wedding Anniversary
with two separate events. On the Saturday we had a family
lunch for about thirty five relations and children including
grown up cousins, all three of our bridesmaids attended as well
as Dave's best man. We held the event in a barn belonging to a
neighbour, which has been converted for functions. My ninety-
year-old aunt was able to come although she seemed rather
confused. Peter, Andrew, Rosemarie and Brenda had arranged it
all so we were able to sit back and relax. Peter had enlarged
photos of our wedding and pinned them around the walls,
which made a real focal point. Peter, Andrew and David made
suitable speeches, the sun shone and it was a really wonderful
occasion. David's sister was over in England from Australia and
stayed with us for part of the time that she was in the U.K. She
enjoyed seeing so many relations in one place as she had to
travel widely to see all the other people that she wanted to see.

The next day, we held 'Afternoon Tea' for all our local friends
and neighbours, again in the barn and entertained about forty
including several children. We had managed to borrow a
'bouncy castle' for them, which kept them occupied. We had a
lovely time and our children once again turned up trumps.
Everyone enjoyed the occasion as the sun shone, the food was
good and the company enjoyable.

59.   The Farm Work Continues...

IT WAS DIFFICULT TO COME BACK DOWN to earth after a weekend like that but we seem to be so busy that, 'we have no time to stand and stare.' The shearer was booked to come to shear David's two hundred ewes. David cleared a good space in a barn and made a 'race' with hurdles leading into a pen. Bob was an older man but Dave had had him last year and liked his style of shearing. He was slower than some of the shearers that David had had in the past but so methodical and careful and Dave could easily keep up with tying up the fleeces as they rolled off the sheep.

David went out before breakfast and gathered the sheep with Bud and Sid who walked them across the fields to the barn. He came home for breakfast and was soon out at the barn to begin. Bob sheared in two hour 'runs' stopping then for ten minutes for a drink and to stretch his back; he was able to shear about 20 per hour, so had them all shorn by four o'clock. David then had to take them back to the field before tidying up the barn and putting all the daggings into a sack. The next day he 'phoned the wool merchant to ask him to come and pick the 'wool sheets' up, each containing a number of fleeces (the lorry has a grab with which the driver can swing the sheets onto the lorry).

It can be some weeks before the woolsacks are collected, as the wool merchant has to wait until he has enough from one area to make a load. At one time David would have had a lorry load just of his wool.

David is busy making hay now. The weather is perfect for it and he's already produced over two thousand bales; last year he managed a total of one thousand bales. As he owns no land of his own he utilizes Adams Farm land, which is mainly in the Upton Valley area. The field nearly opposite the Pumping Station yielded about three times as much as the total amount baled last year as it was a particularly bad year. The grass is first cut before it's turned, rowed up then baled; the noise of the baler is very distinctive as it picks up the hay and processes it into rectangular shaped bales and spits them out behind. David then has the job of picking up the bales with a machine called a bale grab with which he can automatically place the bales on a trailer.

He had arranged to bale some straw as well as our 'horsey' friends need it for the loose boxes. Unfortunately he turned it after the combine, rowed it up and it rained and rained and the straw never did dry out sufficiently to bale. I think the 'horsey people have had to get it from a different source.

He had a consultation with the oncologist in August after having a chest X-Ray. The consultant brought up that days X-Ray on the screen together with the one that David had taken initially.

The difference in the two pictures was quite amazing with a marked reduction in size of the tumour! The consultant had spoken when David had had the previous consultation about a drug trial for which Dave might be suitable but he didn't even mention it. David has to see him in December, he's put on the weight that he had lost and feels really fit again. Although he has altered his sheep keeping by selling his ewes, he's having young sheep from Kent to keep for a year when they'll return to their owners ready for breeding. It will cut down the work load and there won't be the stress of lambing although he's retained about thirty odd ewes, including Bridget O'Sullivan of course.

We are very grateful to God and conscious of the prayers of our friends worldwide for Dave's healing this far and are confident that God holds our lives in His hand.

~ End ~